Design and Advanced Robust Chassis Dynamics Control for X-by-Wire Unmanned Ground Vehicle

Synthesis Lectures on Advances in Automotive Technology

Editor

Amir Khajepour, *University of Waterloo*

The automotive industry has entered a transformational period that will see an unprecedented evolution in the technological capabilities of vehicles. Significant advances in new manufacturing techniques, low-cost sensors, high processing power, and ubiquitous real-time access to information mean that vehicles are rapidly changing and growing in complexity. These new technologies—including the inevitable evolution toward autonomous vehicles—will ultimately deliver substantial benefits to drivers, passengers, and the environment. Synthesis Lectures on Advances in Automotive Technology Series is intended to introduce such new transformational technologies in the automotive industry to its readers.

Design and Advanced Robust Chassis Dynamics Control for X-by-Wire Unmanned Ground Vehicle
Jun NI, Jibin Hu, and Changle Xiang
November 2017

Electrification of Heavy-Duty Construction Vehicles
Hong Wang, Yanjun Huang, Amir Khajepour, and Chuan Hu
November 2017

Design and Advanced Robust Chassis Dynamics Control for X-by-Wire Unmanned Ground Vehicle
Jun NI, Jibin Hu, and Changle Xiang

ISBN: 978-3-031-00368-4   print
ISBN: 978-3-031-01496-3   ebook
ISBN: 978-3-031-00001-0   hardcover

DOI 10.1007/978-3-031-01496-3

A Publication in the Springer series
SYNTHESIS LECTURES ON ADVANCES IN AUTOMOTIVE TECHNOLOGY #2
Series Editor: Amir Khajepour, University of Waterloo

Series ISSN: [COMING] Print [COMING]Electronic

Design and Advanced Robust Chassis Dynamics Control for X-by-Wire Unmanned Ground Vehicle

Jun NI
Jibin Hu
Changle Xiang
Beijing Institute of Technology

SYNTHESIS LECTURES ON ADVANCES IN AUTOMOTIVE TECHNOLOGY #2

ABSTRACT

X-by-wire Unmanned Ground Vehicles (UGVs) have been attracting increased attention for various civilian or military applications. The x-by-wire techniques (drive-by-wire, steer-by-wire, and brake-by-wire techniques) provide the possibility of achieving novel vehicle design and advanced dynamics control, which can significantly improve the overall performance, maneuverability, and mobility of the UGVs. However, there are few full x-by-wire UGVs prototype models reported in the world. Therefore, there is no book that can fully describe the design, configuration, and dynamics control approach of full x-by-wire UGVs, which makes it difficult for readers to study this hot and interesting topic.

In this book, we use a full x-by-wire UGV, developed by our group, as the example. This UGV is completely x-by-wire with four in-wheel motors driven and a four-wheel independent steer steer. In this book, the overall design of the UGV, the design of the key subsystems (battery pack system, in-wheel motor-driven system, independent steer system, remote and autonomous control system), and the dynamics control approach will be introduced in detail, and the experiment's results will be provided to validate the proposed dynamics control approach.

KEYWORDS

unmanned ground vehicle, x-by-wire, vehicle design, dynamics control, autonomous control, in-wheel motor driven, independent steer

Contents

Preface

In recent years, Unmanned Ground Vehicles (UGVs) have been rapidly developing for both military and civilian applications. With no human on board, UGVs are supposed to replace humans for a variety of applications, such as agricultural irrigation, logistics, or express delivery in civilian applications, and reconnaissance, rescue, search, or combat in military applications. It has been widely accepted that the UGVs are able to greatly change transportation, human lives, and the form of land war in the future. Therefore, the research of the UGVs has attracted a great deal of attention worldwide. In addition, the x-by-wire techniques (drive-by-wire, steer-by-wire, and brake-by-wire techniques) provide the possibility of achieving novel vehicle design and advanced dynamics control, which can significantly improve the overall performance, maneuverability, and mobility of the UGVs.

This book focuses on this hot and interesting topic, x-by-wire UGV. In this book, a full x-by-wire UGV, which is developed by the authors, is taken as the example to show the concept, design, control, and functions of x-by-wire UGV. The name of the UGV proposed in this book is Unmanned Ground Carrier (UGC), which is a mother-child and full x-by-wire heavy-class UGV. The details of the design of the UGC is described in this book, including the design of the hardware configuration, signal flow, battery pack, in-wheel motor driven system, independent steer system, and other systems. Based on the full x-by-wire and four-wheel independently actuated technique, the advanced chassis dynamics control is described, including the different independent steer modes, robust H infinity yaw moment control, tire traction force distribution control, and the tire slip ratio control. In the last chapter of this book, the experiments results are provided to evaluate the performance of the UGC.

The authors want to thank to Prof. Amir Khajepour at the University of Waterloo, who was very helpful with the publishing of this book. In addition, the authors want to thank Ph.D. students Yue Zhao, Naisi Zhang, Bo Pan, Hanqing Tian, and Yunxiao Li who made great contributions in the development of the UGC as well as to the organization of this book.

Jun NI, Jibin Hu, and Changle Xiang

CHAPTER 1

Unmanned Ground Vehicles: An Introduction

1.1 BASIC INTRODUCTION TO THE UGV

1.1.1 WHAT IS A UGV?

In recent years, with the rapid development of intelligent techniques, sensor techniques, and vehicle control techniques, Unmanned Ground Vehicles (UGVs) are rapidly developing for both military and civilian applications. The main feature of the UGV is that it operates with no onboard human presence. Therefore, UGVs are usually developed for special applications usage. Through remote or autonomous operation modes, UGVs are able to replace humans for a variety of applications, such as agricultural irrigation, logistics, or express delivery in civilian applications, and reconnaissance, rescue, search, or combat in military applications.

Essentially, the typical functions of a UGV can be divided into the following three functions: "Understand where I am going and what should I do," "Understand the environment and find a path through," and "Control the actuators of the vehicle to follow the desired path." These three parts are just like the brains, eyes, and legs of a human being. These three functions of the UGVs are achieved by corresponding hardware and software systems. In order to achieve the three functions mentioned above, the software systems of an UGV can be divided into four parts: Mission Decision and Navigation Module; Environment Perception Module; Path Planning and Localization Module; and Chassis Dynamics Control Module.

The Mission Decision and Navigation Module understands the mission and calculates the global path planning for the UGV. The global path should be determined before the start of the mission. In addition, the global path planning also greatly relates to the categories of the missions. Take the reconnoiter mission as example: the UGV is supposed to travel through and search the whole area in the reconnoiter mission. Therefore, the shortest global path for searching the whole area should be calculated by the Mission Decision Module, which will be used to guide the UGV to finish the mission. The Environment Perception Module calculates and outputs the environment information for path planning. Generally, the environment information can be obtained using two approaches. The first approach uses the equipped sensors, including the cameras, lidars, or infrared equipment, to collect and understand the environment information. The second approach uses the information provided by the communication network, such as the information provided by the connected vehicle system. The Path Planning and Localization Module is used to find a desirable path, which avoids all the obstacles for the UGV to achieve the target or finish the mission, with consideration of the time or the power consumption. The last part is the Chassis Dynamics Control Module, on which this book is primarily focused. According to the calculated desirable path and vehicle speed, the Chassis Dynamics Control Module should control all the actuators (drive, brake, or steer actuators) of the UGV to achieve the target speed and finish the desirable path with desirable handling stability and dynamics behavior.

1.1.2   EXAMPLE OF UGVS DEVELOPED BY STANFORD UNIVERSITY AND BEIJING INSTITUTE OF TECHNOLOGY

In order to show the basic configuration and software system of a UGV, two typical UGVs are shown as examples here. The first is a very famous UGV-Junior, which was developed by Stanford University. The other one is an autonomous racecar, which was developed by the authors (Beijing Institute of Technology, China).

The Junior is modified based on a 2006 Volkswagen Passat Wagon [1]. The following sections will discuss that the modification based on an existing chassis products is a common approach to manufacture a UGV. In order to add the "brains" and "eyes" on the Junior, five lidars (manufactured by IBEO, Riegl, SICK, and Velodyne), one Global Positioning System (GPS)/Inertial Navigation System (INS), BOSCH radars, and a computer are equipped. Besides the UGV modified based on a passenger car, the autonomous racecar shown in Figure 1.2 is another typical category. The autonomous racecar is modified based on an electric Formula Student racecar, which was developed by the authors. In Figure 1.2, the body covered is removed to clearly show the components. It can be seen that a lidar, a camera, a computer, a GPS/INS, steer servomotors, and brake servomotors are equipped to modify it to autonomous mode.

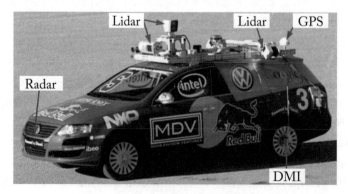

Figure 1.1: Junior by Stanford.

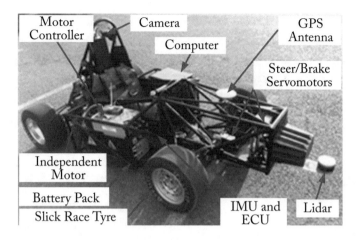

Figure 1.2: Autonomous racecar by Beijing Institute of Technology (BIT).

Strictly speaking, the Junior is not a UGV since it is modified based on a passenger car, which allows people to be on board. However, it can be used to understand the general software system of a UGV. Figure 1.3 shows the overall architecture of the software, which can be organized into five modules. The first module is the sensor interface, which consists of the external sensors, such as the lidars, and makes the sensor data available to the rest of the software based on the data fusion technique. The second module is the perception module, which uses the environment data and information to calculate the static and dynamic obstacles around the vehicle. The third is the navigation module, which consists of a group of path planners, as well as the steer/throttle/brake controllers. The fourth is the drive-by-wire interface, which enables software control of the throttle, brake, steer, gear shifting, and emergency brake. Finally, the fifth is global services, which consists of a number of system-level modules and provides time stamping, logging, and watch dog functions to keep the whole software running reliably.

According to Figure 1.4, the control architecture of the autonomous racecar can be divided into environment sensor system, control system, power system, actuator system, and data acquisition system. The environment sensor system includes a lidar and a camera, which are used to recognize the locations and the colors of the cones. The lidar is mounted in the front of the racecar, and the camera is mounted on the top of the racecar. The control system consists of a computer and a raid Electronic Control Unit (ECU), which is operating in a Robot Operating System (ROS) and Simulink environments, respectively. The computer calculates the environment information and outputs the path planning results. Based on the desired path, the rapid ECU conducts the path following and dynamics control by calculating the desired steer angle or desired traction/brake force. The signal between the computer and the ECU is transferred by a Controller Area Network (CAN) card, which transfers the RS232 signal to the CAN signal. The output of the ECU is a CAN signal, which can be directly used by the actuators. The actuator system includes two independent driven motors—a steer servomotor and a brake servomotor. Each wheel of the rear axle of the racecar is independently driven by a motor. Finally, the power system of the autonomous racecar includes a high-voltage battery pack and DC/DC system. The independent motors' controllers are powered by the high-voltage battery pack, and the low-voltage system is powered by the DC/DC system. The Battery Management System (BMS) is equipped to monitor the battery pack's status, such as the State of Charge (SOC).

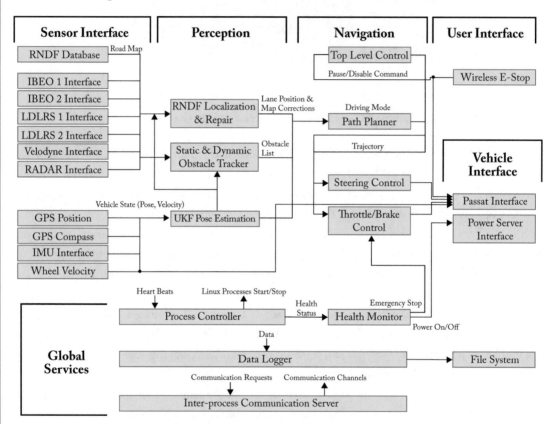

Figure 1.3: Overall architecture of the software of the Junior [1].

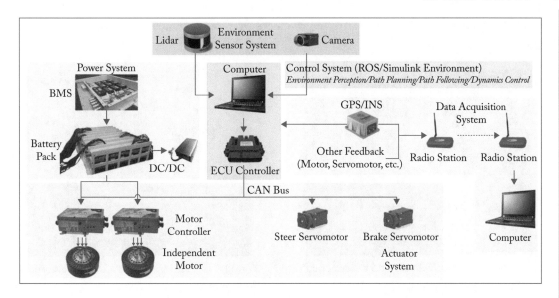

Figure 1.4: Overall architecture of the software of the autonomous racecar.

1.1.3 BASIC APPLICATION OF THE UGV IN CIVILIAN AND MILITARY FIELDS

Now that a basic introduction of UGVs has been provided using the previous examples, the application of the UGV in both civilian and military fields can be discussed. In the civilian field, it has been widely accepted that the emergence of UGV can significantly reduce the labor cost and improve the working efficiency. Due to the technique and law issues, actually, UGVs have not been widely used in civilian road applications. However, there is no doubt that they will be widely used and greatly change human's life in the future. Currently in China, a major application of the civilian UGV is the unmanned express delivery vehicle. As is widely known, the express industry is very hot in China now due to the rapid development of electronic commerce. The average number of the packages per day exceeds one hundred million! Apparently, the labor cost for the express delivery man is very high [2]. Therefore, the application of express delivery UGV is very helpful to reduce labor costs and improve delivery efficiency. However, due to the law issues, these express delivery UGVs are currently used in the closed field, such as the campus or the living garden. Figure 1.5 shows the express delivery UGVs developed by the Jing Dong company [3], which is a famous electronic commerce company in China. This UGV is modified by an electric vehicle, equipped with lidar and cameras. It can support 7–10 packages at one time, and is used for delivering packages in the closed field, such as a university campus. Recently, UGVs have been applied in some universities in China. Apparently, the express delivery UGV is a desirable application of the UGV in the civilian field, which has also attracted the focus from many companies in the world, such as Amazon and Yelp [4].

Figure 1.5: Express Delivery UGV developed by Jing Dong Company in China [3].

Figure 1.6: Express Delivery UGV developed by Yelp in the U.S. [4].

In the military field, UGVs are also attracting great attention due to their unique advantages. it has been widely accepted that the emergence of UGVs is going to greatly change the form of the land war in the future. Many countries, such as China, the U.S., Russia, Germany, and England, are focusing on the development of military UGVs for the applications of reconnaissance, rescue, search, and/or even combat in both urban and field environments. Generally, the application of the UGV in the military field has the following major advantages.

1. With weapons equipped, the military UGVs are able to go into combat without humans on board. Therefore, the soldier's life can be saved. In addition, since there is no human on board, many complex or difficult missions can be conducted, such as long-time reconnaissance and airdrop.

2. The chassis of the UGVs are usually x-by-wire to provide convenience for the sensor system and control system design. With x-by-wire techniques, such as independent electric motor drive techniques, the overall performance of the chassis can be significantly improved, especially the mobility and maneuverability in field environments.

3. Based on the functions of environment reorganization, target tracking, localization, guidance, and wireless long-distance communication, a UGV can be considered an important element in whole military unmanned systems, in collaboration with the Unmanned Aerial Vehicles (UAVs) and Unmanned Underwater Vehicles (UUVs).

Figure 1.7: GXV Concept [5] proposed by the Defense Advanced Research Projects Agency (DARPAR) in the U.S.

In order to further understand the application of military UGVs, several concept examples proposed by different countries will be shown. Figure 1.7 shows the GXV concept proposed by the DARPAR in the U.S. The GXV is supposed to be a light-class four-wheeled UGV with high mobility. As Figure 1.7(a) shows, it can be dropped by the helicopter to achieve the target area and conduct the pre-defined missions, such as search, rescue, or target destruction. Based on its significant mobility, it can move with high speed in the field environments. As DARPAR claims, it is supposed to negotiate 95% of the terrains on Earth. Figure 1.7(b–d) shows serval examples proposed by DARPAR's video.

Figure 1.8 shows the GCV concept proposed by the Russian Army. Different than the GXV proposed by the DARPAR, the GCV is supposed to work in the urban environments. As Figure 1.8 shows, the chassis of the GCV is relatively low, which means it is not able to perform well in the field environments. However, it has great work performance in the urban environments. According to the video proposed by the Russian Army, the GCV is equipped with active suspension and all-wheel-independent-steered techniques. Therefore, the GCV can be steered by the skid steer approach or zero radius steer approach as Figure 1.8(a) shows, which enables the GCV to negotiate the narrow spaces in urban environments. Figure 1.8(a) also shows that the GCV is operated by a human operator through the command station, with the feedback images. Figure 1.8(b) shows that the GCV can be transferred by a larger military transportation vehicle to achieve the target area. With weapons onboard, the GCV can conduct combat missions. The video proposed by the Russian Army shows some examples of how the GCV performs combat in a city.

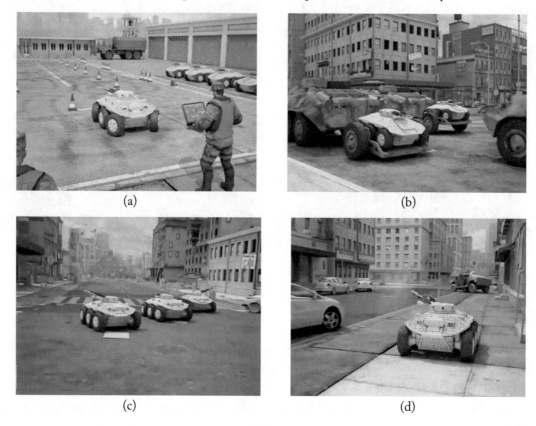

Figure 1.8: GCV concept [6] proposed by the Russian army.

In this section, the basic architecture and functions of a UGV are discussed, and some examples of a UGV in both civilian and military applications are shown. The techniques involved in a UGV is very complex, such as artificial intelligence, environment perception, data fusion, path and motion plan, vehicle dynamics, and control. This book focuses on the overall design and dynamics control of the UGV. The military UGV should receive more focus due to its more flexible configuration and higher overall performance requirement. In next section, the development history of the military UGV will be briefly introduced.

1.2   THE BRIEF DEVELOPMENT HISTORY OF THE MILITARY UGV

1.2.1   LIGHT-CLASS MILITARY ROBOTS

As the previous section introduced, the UGV has attracted a lot of research focus in the military field due to its great advantages. Currently, the existing military UGVs can be divided into two major categories: light-class robots and heavy-class UGVs. Figure 1.9 shows several light-class robots, developed in China, including combat robot, search robot, bomb manipulate robot, and climb robot. The light-class robots can be divided into tracked type or wheeled type. Due to their abilities and potentials to perform complex maneuvers, the light-class robots have been widely used in military applications. The development of various novel locomotion mechanisms, such as track mechanism, omnidirectional steer mechanism, spherical wheels, and transformable wheels, has provided great possibilities for light-class mobile robots to improve their mobility and maneuverability. In military applications, the light-class mobile robots are mainly developed for auxiliary missions, such as search, bomb disposal, or sample collection due to their small size, low locomotion speed, and low carrying load capacity.

(a) Combat Robot [7]

(b) Search Robot [7]

(c) Bomb Manipulate Robot [8]

(d) Climb Robot [7]

Figure 1.9: Several light-class robots in China.

However, in recent years, the demand of applying UGVs in real battle scenarios has increased. It is hopeful that military UGVs will be able to conduct more complex missions, such as border patrol, carrying equipment to reduce solders' carrying load, long-distance transportation, or combat with weapons equipped. Currently, the light-class robots are not able to conduct these missions due to their small size, low speed, and low load capacity. Therefore, it is necessary to develop heavy-class UGVs with high speed, high endurance, and high load capacity, a current worldwide focus.

1.2.2   MILITARY HEAVY-CLASS UGVS BASED ON MODIFICATION

Based on the experience in the intelligent technique, sensor technique, environment perception technique, and motion plan technique in the robotic field, researchers are able to rapidly develop heavy-class military UGVs.

(a) Volkswagen (Stanford University, U.S.) [1]

(b) Land Rover (MIT, U.S.) [9]

(c) General Motors (Carnegie Mellon
University, U.S.) [8]

(d) Volkswagen (University of Karlsruhe,
Germany) [11]

Figure 1.10: Modified UGVs in DARPA Urban Challenge.

Most of the reported heavy-class military UGVs are modified based on existing vehicle products. Therefore, the developers can concentrate more on the control and sensory systems to reduce development cycle and cost. As Section 1.1 introduced, the most successful examples are the UGVs developed for the DARPA Grand Challenge in the U.S. Most of the UGVs developed for this challenge are modified based on passenger car products, such as Volkswagen (Stanford University), Land Rover (MIT), General Motors (Carnegie Mellon University), and Volkswagen (University of Karlsruhe). In the early stage of the heavy-class military UGVs development, the researchers placed more focus on the development of control and sensory systems with adding additional sensors, such as lidars, radars, or cameras. Therefore, it is the most convenient way to use the existing vehicle products to develop a UGV based on modification. Through the x-by-wire technique, the vehicle can be easily modified to be autonomously manipulated by adding servomotors for steer column, throttle, and brake pedals.

Since 2009, the Intelligent Vehicle Future Challenge (IVFC) has been held in China, supported by the Natural Science Foundation of China (NSFC). In 2016, the participant teams have exceeded over 20, and most of them are from universities. This challenge makes great contributions for the development of UGVs in China. Figure 1.11 shows the participant vehicles from Qinghua and Xian Jiaotong University. All of the participanting vehicles in this challenge are also modified based on existing vehicle products, like the vehicles in the DARPA urban challenge. Researchers in China also put their focus on the sensory and control systems development. Therefore, they also used the existing vehicle products to modify the vehicle. Besides the IVFC held by NSFC, there is another important challenge of UGV in China—the "Overcome Danger" Challenge, which is held by the General Armament Department of People Liberation Army (PLA) of China. Since this challenge is held by the military, it placed more focus on the performance of the UGV in field environments, which is very important for a military vehicle. The tasks of "Overcome Danger" Challenge include various missions to test the ability of the UGVs to autonomously or semi-autonomously achieve the target in the field environments, especially the ability to pass the natural and artificial obstacles, such as sand, swamps, broken walls, and moats. Therefore, the major purpose of this challenge is to test the performance of the UGV chassis, especially its driving performance and mobility, which is very different than that of the IVFC. Therefore, most of the participant UGVs are modified from military vehicles because of their higher driving performance and mobility. Two examples are shown in Figure 1.12.

(a) Qinghua University

(b) Xian Jiaotong University

Figure 1.11: Modified UGVs in the Intelligent Vehicle Future Challenge in China.

(a) Inner Mongolia Machine Group

(b) North Research Institute of Military Vehicle

Figure 1.12: Modified UGVs based on military vehicles in the "Overcome Danger" Challenge in China.

However, although the modification approach to developing UGVs enables the researchers to reduce the development cycle and put more concentrate on the sensory systems, it still has apparent drawbacks.

1. The overall configuration of the UGV is strictly limited by the original vehicle chassis, such as the position of the sensory systems, computer system, and battery system. In addition, the original actuator or the steer/brake/throttle is not drive-by-wire, and the servomotors should be added to manipulate the steer column and brake/throttle pedals.

2. Due to the lack of the x-by-wire, it will be difficult for the designers to design and add the control and sensory systems. In addition, the overall performance of the UGV, such as the controllability, mobility, and maneuverability, will also be limited due to the lack of the x-by-wire technique.

Therefore, the modification based on the vehicle chassis product is just a temporary way to develop the UGV. In the particular stage of the UGV development, this way enables the researchers to quickly put their focus on the environment perception and motion plan technique. However, due to the increasing requirements of UGV performance, the x-by-wire technique should be more widely used for UGV, especially for the military UGV, which requires higher performance in both urban and field environments.

1.3   THE DEVELOPMENT OF THE X-BY-WIRE MILITARY UGV

1.3.1   THE SIX-WHEELED SKID-STEERED X-BY-WIRE UGV

Recently, the development of x-by-wire technique (drive-by-wire, steer-by-wire, and brake-by-wire) has provided possibilities to significantly improve the configuration flexibility and overall performance of the ground vehicle. For example, the drive-by-wire technique, especially the four-wheel-independent-drive and torque distribution techniques, can significantly improve the driving performance of the vehicle. Numerous research has been reported in the passenger car field. The involvement of the x-by-wire technique in military UGV has the following advantages.

1. The overall configuration flexibility of the UGV can be significantly improved. The traditional vehicle is steered by the steer column and steer linkages, and powered by the engine and transmission system. For example, if the full x-by-wire technique is applied and all the wheels are driven/steered by the motors, all the mechanical connection will be canceled. Therefore, the overall configuration flexibility can be greatly improved, which is important for developing the new-configured and future military UGV.

2. The involvement of x-by-wire technique also greatly improves the convenience of designing the control and sensory system. For example, if all the wheels are driven/steered by the motors, all the motors will be controlled by the ECU controller through CAN bus. Therefore, the other control and sensory system, such as the computer system for path planning and the sensor system for the environment perception, can be easily connected to the ECU controller, which provide great convenience.

3. The overall performance of the UGV, such as the driving performance, handling performance, mobility, and maneuverability, will be greatly enhanced due to the involvement of the x-by-wire technique. For example, the driving performance will be greatly improved if the all-wheel-independent-drive technique is applied. The mobility in narrow space will be greatly improved if the all-wheel-independent-steered technique

is applied. In addition, the advanced dynamics control can be easily achieved to further improve the overall performance of the UGV.

Due to the above great advantages, the x-by-wire heavy-class military UGVs emerging in recent years are designed totally according to the particular requirement of UGVs, rather than just modifying it based on existing vehicles. The major categories of x-by-wire heavy-class military UGVs include six-wheeled UGVs and four-wheeled UGVs.

(a) MULE (Lockheed Martin)

(b) Crusher (Carnegie Mellon University)

Figure 1.13: Six-wheeled x-by-wire UGVs [12].

The Multifunctional Utility/Logistics and Equipment Vehicle (MULE), developed by Lockheed Martin, and the Crusher, developed by Carnegie Mellon University, are two of the most famous six-wheeled x-by-wire military UGVs [12]. The total weight of the MULE is 2500 kg, and it is featured as a common chassis to conduct different missions with different equipment. The MULE chassis is equipped with six independent in-wheel motors. Therefore, the volume in the body can be significantly saved to locate other equipment. Through torque force distribution and tire slip ratio control technique, the driving performance of the MULE can be greatly improved. In addition, there is no particular steer mechanism. Therefore, the MULE is steered by the skid-steered technique. The MULE is equipped with six independent articulated suspensions, which provide great mobility to negotiate complex terrain and obstacles. As a common chassis, the MULE has several variants, including XM1217, XM1218, and XM1219, which are equipped with different equipment to conduct different missions, such as detect, mark, reconnaissance, surveillance, and combat [12]. Although the missions to develop the MULE has been canceled, the emergence of the MULE series is still a landmark in this field. The Crusher is another famous six-wheeled x-by-wire UGV, which is developed by Carnegie Mellon University. The weight of the Crusher is 6000 kg, which is much higher than that of MULE. The crusher is also driven by six in-wheeled motors, and it is also steered by skid-steered technique without any particular steer mechanism. Since the Crusher is developed by the university, it has been applied in the military. Both the MULE and Crusher are powered by a hybrid electric system, which enables them to operate mutely (powered only by battery).

1.3.2   A SIX-WHEELED UGV DEVELOPED BY THE AUTHORS

(a) Overview

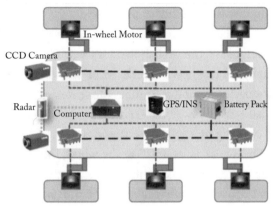

(b) Overall Configuration

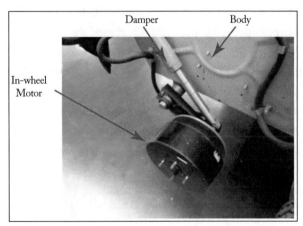

(c) Details of the In-wheel Motor

Figure 1.14: Six-wheeled x-by-wire UGV developed by our group [13].

In order to further understand the configuration of the six-wheeled x-by-wire UGV, a testbed UGV developed by our group at Beijing Institute of Technology is shown in Figure 1.14 [13], as well as the illustration of the overall configuration. The vehicle specifications are shown in Table 1.1. The mass of the UGV is only 800 kg, since this UGV is just a prototype vehicle. The body length is 3 m and the track width is 1.6 m. It can be seen that the vertical load of the 1st axle is 250 kg, the vertical load of the 2nd axle is 400 kg, and the vertical load of the 3rd axle is 150 kg.

The vertical load of the 2nd axle is the heaviest, which is a unique feature of the six-wheeled UGV. The reason is because the six-wheeled UGV is steered by the skid-steered technique, the steering resistance force in pivot steer condition is mainly caused by the front and rear tires. Therefore, if the vertical load on front and rear tires are designed to be smaller, and larger vertical load is designed on the middle tires, the steering resistance force will be significantly reduced, which can greatly improve the mobility of the vehicle.

The UGV shown in Figure 1.14 is powered by a high-voltage battery pack, which is different than that of the MULE and Crusher. This UGV is also driven by six in-wheel motors, which are equipped inside the wheels, while six motor controllers are mounted inside the body frame and connected to motors by three-phase electric wires. The battery pack and the computer are mounted inside the body frame. The CCD cameras and radars are mounted at the front of the vehicle. According to Figure 1.14, the advantage of the x-by-wire UGV can be easily seen. Since each in-wheel motor is mounted in the wheel, the volume of the body can be significantly improved and the configuration flexibility can be greatly improved. The overall architecture of the path following and dynamics control module is shown in Figure 1.15, which can help the readers to understand how the chassis dynamics control works for a six-wheeled x-by-wire UGV. The details of the environment perception and path-planning modules are neglected. The desired speed and desired path of the UGV is determined by the path-planning module. Therefore, the tracking module aims at tracking the desired speed and desired path of the UGV based on a PID controller and a path-tracking algorithm. The speed PID controller calculates the desired longitudinal acceleration and traction force, and the path tracking algorithm calculates the desired yaw moment based on a reference dynamic model according to the desired yaw velocity. As a six-wheel-independent-drive vehicle, the most important is how to distribute the traction force of six wheels. Therefore, a longitudinal tire force distribution module is used to calculate each motor's traction force. The details can be seen in the authors' previous work [13].

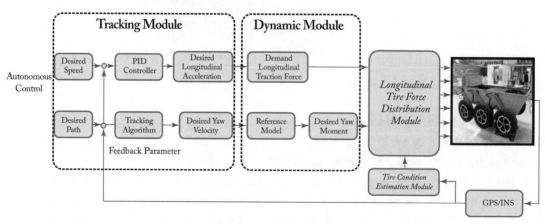

Figure 1.15: Overall architecture of dynamics control of six-wheeled x-by-wire UGV.

Table 1.1. The vehicle specification of the six-wheeled x-by-wire UGV

Parameter	Value
Mass	800 kg
Body height	1 m
Body length	3 m
Track width	1.6 m
Moment inertia around z axis	550 kg·m²
Distance from 1st axle to C.G	0.5 m
Distance from 2nd axle to C.G	0.25 m
Distance from 3rd axle to C.G	1 m
Vertical load of 1st axle	250 kg
Vertical load of 2nd axle	400 kg
Vertical load of 3rd axle	150 kg
Height og C.G	0.8 m
Stiffness of suspension spring	50 N/mm
Radius of wheel	0.3 m
Max power of each motor	12 kW

1.3.3   THE FOUR-WHEELED X-BY-WIRE UGV

As previously introduced, the full x-by-wire technique was first used for six-wheeled skid-steered UGVs. As is widely known, the load capacity is a very important index for the military vehicle. Actually, the load capacity of six-wheeled UGV is much higher than that of a four-wheeled UGV, which means it can support more equipment or materials in the missions. However, the six-wheeled UGV still has apparent drawbacks.

1. The six-wheeled UGV is steered by the skid-steered technique, which means that all the particular steer mechanisms can be omitted, and the vehicle can be steered by the traction force difference between each side's wheels. However, although the Ackermann steered mechanism can be omitted to improve the configuration flexibility, many drawbacks still emerge. At first, since the skid steered technique uses the force and speed difference between each side's wheels to steer the vehicle, therefore, the large tire skid and slip is unavoidable. Consequently, the tires' life will be greatly reduced. As is widely known, a tire's life is very important for a wheeled military vehicle.

2. In addition, the handling performance and the path tracking performance will also be greatly degenerated due to the application of the skid-steered technique. As is widely known, for an Ackermann steered vehicle, the steering trajectory is constrained by the Ackermann principle. However, for the skid-steered vehicle, the vehicle steering trajectory can not be constrained by any kinematics principle, since it uses speed and force difference to steer the vehicle. Consequently, the handling performance of the skid-steered UGV will be degenerated. Also, the path tracking performance, which is important for an UGV, will also be degenerated. In our previous work [13], this drawback has been theoretically and experimentally proved based on the six-wheeled UGV tested shown in Figure 1.14.

Therefore, it has been widely accepted that the six-wheeled UGV is not appropriate for the application in urban environments due to its bad handling and path tracking performance. In the view of practice, the amount of combat possible in an urban environment is rapidly increasing in recent years, such as the combat against terrorism or other small-scale combats, which requires the military UGVs should have good performance in both field and urban environments. In urban environments, the UGV must have good handling and path tracking performance to perform well on the pavement road. Apparently, although the skid-steered six-wheeled UGV can perform well on uneven roads, it cannot perform well on the pavement road because of its particular steer mechanism. Based on the above reasons, the four-wheeled UGV has better potential and balance to have great handling performance in both field and urban environments. Therefore, the four-wheeled x-by-wire UGV has attracted increased focus in recent years.

(a) MDARS   (b) Guardium

Figure 1.16: Famous four-wheeled military UGV [12].

Figure 1.16 shows two famous four-wheeled military UGVs before the emergence of the x-by-wire UGV. As the picture shows, the small-size wheels enable the UGV to perform well in urban environments. Take Guardium as an example. It is modified based on an All-Terrain Vehicle (ATV) TomCar chassis. The Guardium is developed by G-NIUS, and is being used by the Israeli Army. It can be used in either remote or autonomous mode and has been reported to have been successfully applied along the border. Based on the modification of the TomCar Chassis, the total weight of the Guardium is 1.4 tons with a highest speed of 80 km/h. Although the Guardium has been proved with good performance and has been successfully used in practice, its overall performance is still much lower than the four-wheeled military vehicle with full x-by-wire technique.

As introduced above, in recent years more researchers and companies have put their focus on the x-by-wire four-wheeled military vehicle. The Gecko developed by the German Army, shown in Figure 1.17, is a famous x-by-wire four-wheeled military vehicle. The Gecko is a typical all-wheel-independently-steered (AWIS) and all-wheel-independently-drive (AWID) UGV with full x-by-wire technique applied. The weight of Gecko is 3,000 kg, which is much larger than that of MDARS and Guardium. With hybrid diesel and electric powertrain, Gecko's each wheel is driven by an in-wheel motor. The highest speed achieves 80 km/h. The driving performance especially in off-road conditions can be significantly improved due to the application of drive-by-wire technique. Moreover, Gecko also adopts the steer-by-wire technique. Each wheel is steered by a servomotor, which provides the possibility for zero radius turning by placing each wheel to a predefined position. Compared to the pivot steer technique of the skid-steered six-wheeled UGV, the zero radius turning can significantly improve the performance and reduce the tire skid. In addition, the handling performance and path tracking performance problem of the skid steered six-wheeled UGV can be solved by the four-wheeled AWIS technique.

For the UGV, the sensors for environment perception, including radar, lidars, cameras, or other autonomous-related apparatus are also important, but only the specifications and techniques of the chassis are discussed here because of the key focus of this book.

Figure 1.17: Four-wheeled x-by-wire UGV (Gecko, Germany).

CHAPTER   2

Design of a Full X-by-Wire UGV-Unmanned Ground Carrier

In Chapter 1, the significance and necessity of applying the full x-by-wire technique in UGV design has been discussed, as well as the advantages of the four-wheeled military UGV. In this section, a full x-by-wire UGV, namely Unmanned Ground Carrier (UGC), which is developed by the authors, will be described. In addition, the key subsystems, such as the high-voltage battery pack, in-wheel motor driven system, independent steer system, and the suspension system are described in detail to provide guidance for the readers.

2.1   SERVAL CLASSIC X-BY-WIRE TESTBEDS IN PASSENGER CAR FIELD

Chapter 1 states that the x-by-wire four-wheeled UGV is being widely applied in the military field. Actually, the x-by-wire technique is also widely focused in the passenger car field. The vehicle dynamics control technique in passenger car field dates back to the Traction Control System (TCS) and Anti-block Brake System (ABS). With the development of the vehicle electronics technique and the increasing requirement of the passenger car's safety performance, more x-by-wire techniques are being applied to the passenger car, such as Active Front-wheel Steer (AFS) system, 4 Wheel Steer (4WS) System, Direct Yaw Control (DYC) system, Active Suspension System (ASS), and integrated vehicle dynamics control system. Due to space limitations, they will not be detailed in this book. The readers can find it in many sources [42, 59].

In recent years, many x-by-wire passenger car testbeds have emerged for the purpose of academic research, especially the testbeds with AWID technique, which is a most advanced and widely focused technique. Before the description of the UGC, several famous and classic x-by-wire passenger car testbeds will be introduced to provide the comparisons. Since there are many common techniques between the four-wheeled x-by-wire passenger car and UGV, it is very necessary to make this comparison.

The x-by-wire passenger car prototype testbeds featured in Figure 2.1 include OIS EV, UOT March II, IMW-EV, and EGV, which are all developed by universities. All of the four-passenger car testbeds are AWID with independent motors driven. The researchers have made great achievements in vehicle dynamics control and vehicle parameters online estimation based on the experiments on these vehicles [14–17]. The OIS EV is designed to demonstrate a novel omnidi

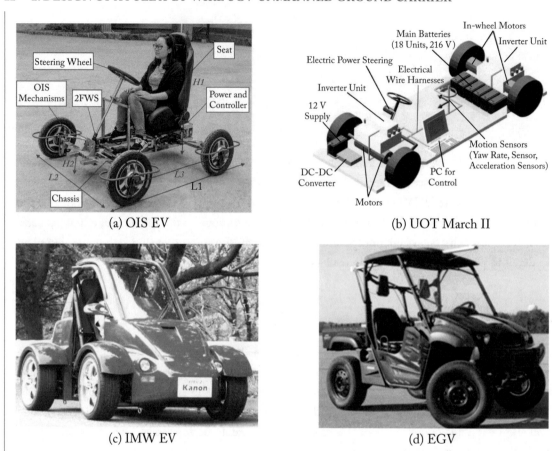

(a) OIS EV

(b) UOT March II

(c) IMW EV

(d) EGV

Figure 2.1: Famous four-wheeled military UGV [14–17].

rectional steer mechanism to achieve AWIS. Therefore, it can be seen that the configuration of the OIS EV is relatively simple. The length of the OIS EV is 1,100 mm, and the width is 1,500 mm. It is powered by pure electric with four DC motors. The maximum power of the OIS EV is only 3.2 kW, which means the maximum power of each independent motor is only 0.8 kW. In Zhang et al. [14], the novelty of the steer mechanism is described in detail. The UOT March II, which is shown in Figure 2.1(b), is very famous in this field, which is modified based on a passenger car chassis. Its total weight is 1,400 kg, and it is also purely electric with four in-wheel motors driven. The steer mechanism of the UOT March II is still a traditional Ackermann steer. The maximum power of the UOT March ii IS 36 kW. Based on the UOT March II testbed, the researchers have successfully made great achievements in the early stage of the vehicle dynamics control of AWID EV. The maximum power of IMW-EV, shown in Figure 2.1(c), is relatively higher at 80 kW, which means the power of the in-wheel motor has been greatly improved with the development of the motor technique. The IMW-EV is manufactured by the same group as UOT March II. This testbed provides agreat base for the research of the DYC vehicle control and the vehicle parameters online estimation. The EGV EV is modified based on an ATV product with good off-road performance. It is also purely electric with four in-wheel motors driven and the maximum power is 30 kW. The academic achievements based on these testbeds will be discussed in a later section. Based on the above discussion, the state of the art of the x-by-wire testbed development in the passenger car academic field can be seen.

2.2   OVERALL CONCEPT OF THE UNMANNED GROUND CARRIER

2.2.1   THE PROPOSAL OF THE OVERALL CONCEPT

Supported by the Advanced Science and Technology Commission (ASTC) of Central Military Commission (CMC) of China and NSFC, the authors developed a novel four-wheeled full x-by-wire UGV, namely, Unmanned Ground Carrier (UGC), which can be applied for both civilian and military applications.

In summary, the UGC has the following major novelties and contributions. (1) As discussed in Chapter 1, the previous UGVs are supposed to be designed as individual vehicles to conduct various missions. For the UGC proposed by the authors, a major novel contribution is applying the "Carrier" concept. In other words, the UGC will no more be an individual vehicle like previous UGVs. The UGC is a mother-child type UGV, which is a platform to carry rotorcrafts or other carrier-based equipment and small robots to improve the overall performance and conduct complex missions through collaborative control. This has strong application potentials in both civilian and military field. (2) In order to improve the performance in both urban and field environments, the mother vehicle of the UGC adopts full x-by-wire and four-wheel-independently-actuated techniques. Each wheel of the mother vehicle is driven by an independent in-wheel motor, and steered by an independent servomotor, which provides great potential for achieving advanced dynamics control to significantly improve the overall performance in both urban and field environments. With independent-steered technique, the mother vehicle can achieve four different steer modes, including Ackermann Steer Mode (ASM), Double Axle Steer Mode (DASM), Diagonal Move Steer Mode (DMSM), and Zero Radius Steer Mode (ZMSM), to deal with different operating conditions. With independent-drive technique, the traction forces of each wheel can be distributed flexibly to negotiate complex terrain, and the active external yaw moment can be applied to improve the handling performance.

(a) Military Version   (b) Civilian Version

Figure 2.2: The unmanned ground carrier.

The military and civilian version UGC are shown in Figure 2.2. Both the military and civilian version UGCs use the same full X-by-wire chassis. Since the full X-by-wire chassis uses the high-power density battery pack, AWID and AIWS technique, it can perform very well in both field and urban environments. Therefore, the same chassis is used for both military and civilian versions. Only the landed carrier-based equipment and body cover are different between two versions. As shown in Figure 2.2(a), the military version is covered by the green armor. In addition, the following carrier-based equipment is carried on a military version UGC: (1) an armed rotorcraft, which can conduct the missions of reconnaissance, search, rescue, and combat. Its loading capacity is 35 kg and the endurance time is 60 min. It is equipped with High Explosive Sqush Head (HESH), tear grenade, optical camera. and infra cameras; (2) a twin duct aircraft with an arm, which can conduct the missions of the grab or manipulate in a narrow closed space; and (3) a loitering munition, which can be launched to 1,500 m high with navigation speed as 70–120 km/h. Its wingspan is 1.5 m and its length is 1.2 m. The mass is 9 kg. It can conduct combat missions in a circle with a diameter of 30 km. However, as shown in Figure 2.2(b), the carrier-based equipment is changed in the civilian version UGC. In the civilian version, six rotorcrafts landed on the platform for the civilian applications, such as power instrument patrol or express delivery. In addition, a small tracked robot is located inside the body. The armed aircraft and the twin duct aircraft, which are carried on the military version UGC, are shown in Figure 2.3.

(a)   (b)

Figure 2.3: (a) Armed aircraft and (b) twin duct aircraft.

2.2.2   THE SOCIAL ACTIVITIES OF THE UGC

The development of the UGC received great attention and the UGC has been widely reported by the press [18–20]. In addition, the UGC has been invited to participate in many great social activities. Figure 2.4 shows some examples of important social activities, in which the UGC participated. Figure 2.4(a) shows the UGC participated in the "Overcome Danger" challenge, which was an important UGV competition held by the PLA of China. The UGC was the only UGV to participate and exhibit in this challenge. Figure 2.4(b) shows the UGC participated in the World Robot Conference, which is a worldwide conference in the robotic field. Figure 2.4(c) shows the UGC being visited by many important leaders in China, such as the Vice President of China and the Vice Minister of the Ministry of Science and Technology of China.

2.2.3   THE ADVANTAGES OF THE UGC COMPARED TO PREVIOUS FOUR-WHEELED UGVS

Before the detail description of the concept of the UGC, the drawbacks of the previous existing four-wheeled military UGVs should be analyzed and discussed.

1. Due to the demand of load capacity, the sizes and weights of the heavy-class military UGVs are relatively large. Take the MULE, Crusher, and Gecko as examples: the lengths of the MULE and Crusher are about 3 m, and the length of the GECKO is about 2.5 m. Therefore, these UGVs are not able to work in many narrow spaces due to the large size, such as pipe, buildings, debris, moat, and forest, which are very common in both field or urban environments. Apparently, this disadvantage limits the application of UGVs in many conditions. Actually, the light-class robots or small-sized rotorcrafts can be used to conduct missions in such narrow spaces. However, in

Figure 2.4: (a) Participation in the "Overcome Danger" Challenge, which is held by the General Armament Department of the People Liberation Army (PLA of China); (b) participation in the World Robot Conference; and (c) visit by the Vice President of China, Vice Minister of Science and Technology of China, and Vice Minister of the Ministry of Public Security of China.

view of practice, there are still many inconveniences to use the small-sized robots or rotorcrafts. At first, the locomotion speed of light-class small-sized robots is relatively low, which limits their ability to conduct the long distance missions. In addition, the human operator has to move very close to the target area to send the robots. Otherwise, it will take too much time for them to achieve the target area, which is unacceptable in military applications. However, in some conditions, it may be too dangerous for the human operator to move too close to the target area, such as in the missions of explosive manipulation or enemy detection. Second, the endurance performance of the small-sized robots and rotorcrafts is rather limited because they are not able to carry heavy batteries due to the limit of the loading capacity. Therefore, they are not able to work for too long a time without human operators or charging equipment around. This also limits their applications in both civilian and military applications.

2. In order to negotiate both field and urban environments, the handling performance, maneuverability, and mobility of four-wheeled military UGVs should be greatly improved. At first, most of the previous four-wheeled military UGVs were modified based on existing chassis products, which have been shown in the above section. These modified UGVs use originally traditional engine and transmission systems, which are not able to flexibly distribute the traction force to negotiate the complex terrain in field environments. In addition, most of the modified UGVs use the traditional Ackermann steer mechanism, which limits the maneuverability and mobility in narrow spaces. The lack of the x-by-wire technique limits the possibility of achieving advanced dynamics control. Second, although some x-by-wire four-wheel UGVs have been developed, such as Gecko shown in Figure 1.17, there is still much space to further improve the x-by-wire percentage and controllability. In addition, the particular academic research on advanced integrated dynamics control of UGVs through x-by-wire technique is seldom reported. Although the integrated dynamics control approaches have attracted much academic research efforts in the civilian autonomous vehicle field, they cannot be applied for military UGVs because of the totally different performance requirements.

To this end, the authors propose the overall concept of the UGC [21], which is detailed as follows.

In order to overcome the first drawback discussed above, the advantages of heavy-class UGVs (high load capacity, high speed, and long endurance) and the advantage of small-sized robots and rotorcrafts (ability to work in narrow space) should be combined together to form a totally novel concept— "Carrier" Concept. Therefore, the UGC will not be an individual UGV, and it will become a mother vehicle in collaboration with rotorcrafts and small-sized robots onboard. The mother vehicle can be used to carry the rotorcrafts and the robots to achieve the target area, and then send them out to conduct the missions. After the missions are completed, the mother vehicle can take them back and return. In this way, the drawbacks that the heavy-class UGVs cannot work in narrow spaces and the small-sized robots cannot work for a long time and distance can be resolved simultaneously.

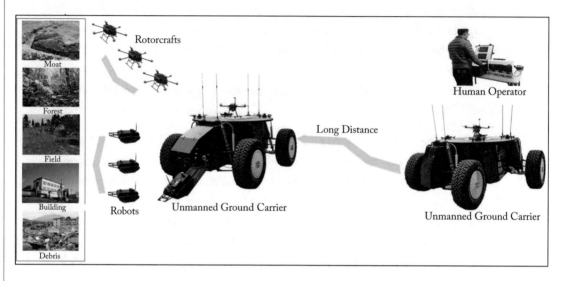

Figure 2.5: The illustration of the concept of UGC.

Figure 2.5 illustrates the scenarios of how UGC works. The human operator can stay in the base, which is far away from the target area. The human operator controls the UGC through a command station, and receives image feedback to determine the next actions. By remote control techniques, the UGC can be controlled to travel a long distance to reach the target area. When the target area arrives, the rotorcrafts and the robots, which are carried by the UGC, will be sent out to conduct the missions in the narrow spaces, such as moats, forests, field, buildings, and debris. According to different types of missions, different rotorcrafts and robots can be selected to be carried on, such as the military version and civilian version shown above. The rotorcrafts and robots are also equipped with cameras to provide feedback images to help the human operator to determine the actions. After the missions are completed, the rotorcrafts and robots can go back to the mother vehicle, and the mother vehicle can go back to the base. In addition, with full x-by-wire, AWID and AWIS techniques applied, the mother vehicle has great mobility to negotiate many kinds of terrains.

Therefore, the requirements of the mother vehicle can be concluded as: (1) high x-by-wire percentage to provide basement for the control, sensory, and communication systems; (2) great maneuverability and mobility to negotiate the terrains in both field and urban environments; and (3) high load capacity to carry the carry-based equipment, such as rotorcrafts or robots.

Therefore, the mother vehicle of the UGC is designed as a full X-by-wire vehicle, which applies electric powered technique, as well as four-wheel-independently-actuated technique. The key subsystems of the mother vehicle include: (1) high-power density battery pack, which aims at improving the endurance and improving the operating time, especially in military use; (2) AWIS mechanism to improve the maneuverability and the mobility; (3) in-wheel-motor-driven mechanism, which aims at achieving AWID significantly improve the driving performance; (4) the mechanism to send out and take back the rotorcrafts and robots; and (5) the command station to control the mother vehicle, rotorcrafts, and robots. In the next section, the details of all the subsystems will be described.

2.3   OVERALL DESIGN OF THE UNMANNED GROUND CARRIER

2.3.1   OVERVIEW OF THE UGC

The UGC was shown in Figure 2.2. The specifications of the mother vehicle are shown in Table 2.1. The vehicle mass of the mother vehicle is 1,200 kg, and the load capacity is 1,000 kg. The wheelbase of the mother vehicle is 2.2 m, and the trackwidth is 1.6 m. The ground clearance is 0.4 m at static, which can be adjusted by the suspension shock absorber. The maximum power is 250 kW. The highest speed is 120 km/h. These specifications will be detailed in following sections. In addition, the overall design of the UGC has been shown in our previous work [21–23], which the readers are encouraged to seek out for further details.

Table 2.1: Specifications of the mother vehicle specifications values vehicle mass 1,200 kg wheelbase

Specifications	Values
Vehicle mass	1,200 kg
Wheelbase	2.2 m
Track width	1.6 m
Height	1.4 m
Tire radius	0.44 m
Ground clearance	0.4 m
Maximum power	250 kW
Battery capacity	15 kWh
Highest speed	120 km/h

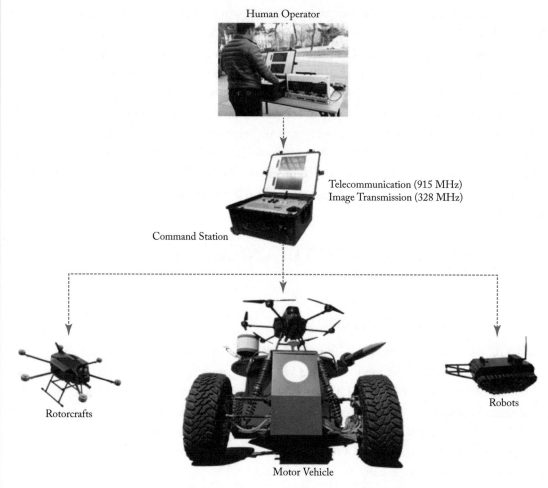

Figure 2.6: Telecommunication of the UGC.

Figure 2.6 shows the telecommunication between the mother vehicle, rotorcrafts, and robot in remote-operated mode. The human operator controls the UGC by the command station, and observe the image feedback by the digital cameras. The mother vehicle, rotorcrafts, and robot carry their own digital cameras to help the human operator to determine the actions. The monitors are assembled in the command station. The telecommunicating frequency is 915 MHz. The transmitter radio station is assembled in the command station, and the receivers are carried by rotorcrafts, mother vehicle, and robot, respectively, which means that they can be controlled, respectively. In the following sections, the details of the command station and the mechanisms for rotorcrafts and robots will be described.

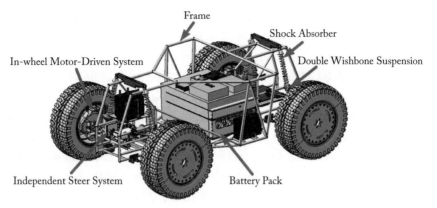

Frame

Shock Absorber

In-wheel Motor-Driven System

Double Wishbone Suspension

Independent Steer System

Battery Pack

(a) Main subsystem of the mother vehicle

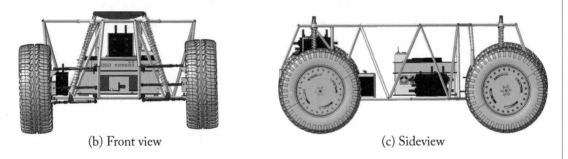

(b) Front view

(c) Sideview

Figure 2.7: Overview of the mother vehicle.

Figure 2.8: Overview of the UGC without body cover.

Figure 2.7 shows the configuration in a 3D model of the mother vehicle, and Figure 2.8 shows the configuration of the whole UGC without body cover. Figure 2.7 shows that the main subsystems of the mother vehicle are in-wheel motor-driven system, independent-steer system, and battery pack, which will be detailed in later sections. There are four in-wheel motor-driven systems and four independent-steer systems to control four wheels. The battery pack is located at the middle of the vehicle. The other instruments are mounted on the top of the battery pack, including DC/DC convertor, radio station, ECU controller, etc. The frame is designed as tube-frame to obtain high torsional stiffness, which is important for working in field environments for military vehicle. The suspension is double wishbone to provide good handling performance. Figure 2.8 shows the platform for the rotorcrafts equipped. The platform is a circle-shaped platform, which is mounted on the body frame by bolts. The rotorcrafts are landed on the platform by the electromagnets, and the details will be shown in a later section. The antennas for communications system are located on the platform. The robot is located at the rear side of the vehicle. The door, which can be open or closed for the robot, is hidden.

2.3.2   THE HARDWARE CONFIGURATION AND THE SIGNAL FLOW OF THE UGC

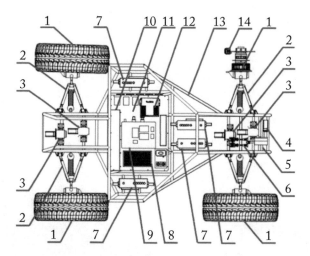

Figure 2.9: Hardware configuration of the mother vehicle.

Figure 2.9 shows the hardware configuration of the chassis of the mother vehicle, and each important component is denoted by a number. The details of each component is shown as follows: (1) in-wheel motor; (2) double wishbone suspension; (3) steer servo motor; (4) brake servo motor; (5) brake gearbox; (6) brake hydraulic cylinder; (7) motor controller; (8) DC/DC convertor; (9)

instruments box; (10) BMS; (11) battery pack; (12) ECU controller; (13) body frame; and (14) brake caliper.

As a unique advantage of the full x-by-wire vehicle, the x-by-wire technique eliminates the mechanical links in the vehicle. Therefore, all the components should be flexibly placed to any position, where they can improve the space saving, wires configuration, assembly convenience, or satisfy other requirements. This advantage can be easily seen in the illustration of Figure 2.9. Actually, the control configured vehicle (CCV) concept is applied during the overall configuration to improve the overall configuration flexibility, which will be detailed in a later section. Figure 2.11 shows the signal flow of the mother vehicle [21], which includes all the important components and the signal flow between them. Through 915 MHz telecommunicating, the control demand from the human operator is received by the radio station carried on the mother vehicle. The radio station outputs RS485 signal, and it is translated to CAN signal and transmitted to the raid ECU controller through a CAN card. As introduced above, each wheel of the mother vehicle is driven, braked, and steered independently by a driven motor or servomotor. Each wheel is independently driven by an in-wheel motor with maximum power as 60 kW. Each wheel is independently steered by a servomotor through a rack-pinion and tie rod linkage. The brake hydraulic cylinder is controlled by a servomotor, and each wheel is braked by a brake caliper, which is mounted on the upright in the wheel. Therefore, the ECU controller controls four PMSM controllers and five servomotors through CAN bus. In addition, a linear servomotor is used to control the door to be open or closed to send out or take back the robot. There are five electromagnets located on the platform, which are used for controlling quadrotors to take off or land. As a pure electric vehicle, the power of the mother vehicle is provided by a 400 V LiFePo4 high-voltage battery pack, which is located in the middle of the frame. The capacity of the battery pack is 15 kWh, which is composited by 432 battery cells. The peak power of the battery pack is 250 kW, which satisfies the demand of the peak power of four PMSMs. Each battery cell is equipped with a Battery Management System (BMS) chip to monitor the battery sates, such as state-of-charge (SOC) or other important parameters of the battery pack. The details of the design of the battery pack will be shown later. Four motor controllers are directly powered by the high-voltage battery pack. By three-phase electric wires connecting, each motor is powered by its controller. All the low-voltage instruments are powered by the DC/DC convertors, including radio station, CAN card, ECU, servomotors, electromagnets, and pumps of cooling system. A 400 V–24 V DC/DC convertor provides the power for the ECU controller and servomotors. A 400 V–12V DC/DC convertor provides the power for the radio station, CAN card, electromagnets, and pumps. Most of the small instruments are located on the top of the battery pack. A differential GPS/INS system is equipped to collect the information of vehicle position and 6-DOF motion data, which is also mounted on the top of the battery pack. The Inertial Measurement Unit (IMU) sensor is located on the top of the battery pack, and two antennas are located at both the front and rear of the vehicle to obtain a longer baseline.

Figure 2.10: Overview of the mother vehicle without body cover.

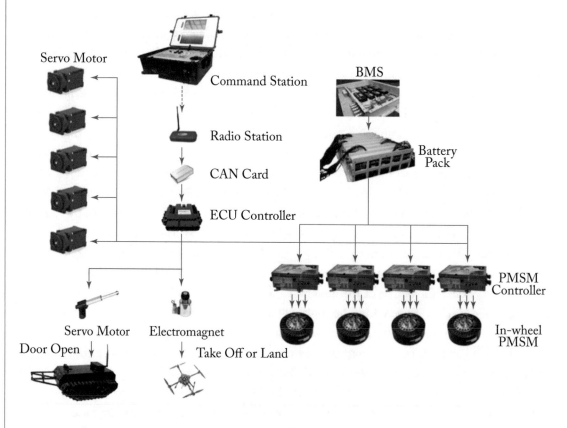

Figure 2.11: Singal flow of the mother vehicle.

2.4   DESIGN OF THE HIGH-VOLTAGE BATTERY PACK AND THE IN-WHEEL MOTOR-DRIVEN SYSTEM

2.4.1   DESIGN OF THE HIGH-VOLTAGE AND HIGH-POWER DENSITY LIFEPO4 BATTERY PACK

The high-voltage battery pack is the most important component of the UGC, which powers the whole vehicle. In addition, all the other low-voltage instruments and sensors are powered by the DC/DCs, which is connected to the battery pack [21–23]. The battery pack is a LiFePo4 battery pack. The voltage of the battery pack is 400 V, and the capacity of the battery pack is 15 kWh, which can supply the mother vehicle to travel 100–200 km. The battery pack is composited by 432 LiFePo4 battery cells. The LiFePo4 battery cell is shown in Figure 2.12(a). In order to obtain the high-power density, the peak discharge rate of the battery cell is designed to be 20 c, which is a very high level. The peak power of the battery pack is 250 kW, which satisfies the demand of the peak power of four PMSMs. The 432 LiFePo4 battery cells are assembled in a 3D printed plastic frame, which is shown in Figures 2.12(b, c). The 3D printed technique make a great contribution for the lightweighting. In order to improve the performance and assure safety of the battery pack, each battery cell is equipped with a BMS chip to monitor the battery sates, such as State of Charge (SOC) or other important concerned states. In addition, the BMS can also balance the battery voltages to increase the life of the battery based on the measured voltage differences among the battery cells. There is no doubt that the strategy of BMS significantly improves the battery's endurance and life. The BMS and all of its components are equipped on the top of the battery pack as Figure 2.12(d) shows. The whole battery pack is located at the middle of the vehicle. The whole weight of the battery pack is only 108 kg due to the application of several light-weigh approaches, such as the application of high discharge rate battery cells and the 3D printed technique. Figure 2.12(e) shows the whole battery pack and the low-voltage instruments on it. Figure 2.12(f) shows the details of the safety circuit and Figure 2.12(g) the up view of the mother vehicle.

(a) Battery Cell

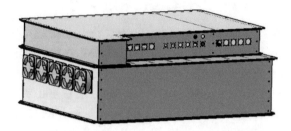

(b) Cells Assembly in 3D Model

(c) Cells Assembled on 3D Printed Frame

(d) BMS

(e) Battery Pack

(f) Safe Circuit

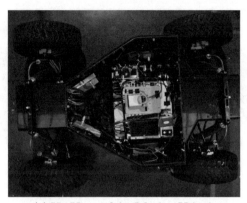

(g) Up View of the Mother Vehicle

Figure 2.12: 400V LiFePo4 battery pack.

2.4.2 DESIGN OF THE COMPACT IN-WHEEL MOTOR-DRIVEN SYSTEM AND PERFORMANCE EVALUATION

The in-wheel motor-driven system is another important component of the UGC. The in-wheel motor-driven technique has two major advantages. First, each wheel can be driven by an independent motor with quick and accurate response, which provides convenience to distribute each wheel's traction force flexibly and achieves tire slip ratio control to negotiate complex terrain. Second, the motor is mounted in the wheel assembly, which significantly saves space in the frame to improve configuration flexibility. However, the in-wheel motor driven technique has not been widely used in spite of the above advantages. The reason is that it is very difficult to develop the mechanism of an in-wheel motor driven system. At first, it places a high requirement for the performance of motor product. The weight of the motor should be as low as possible to reduce the unsprung mass, and the size should be small enough to be mounted in the wheel. Second, the ride performance of the vehicle will be degenerated due to the increase of unsprung mass. The dynamic load acting on the wheel will increase due to the increase of unsprung mass, which may damage the components in the wheel. Therefore, the mechanism design of the in-wheel motor system should be designed to reduce the dynamic load generated by the road roughness.

The key components of the in-wheel motor-driven system are the in-wheel motor and motor controllers. There are four in-wheel motors and four controllers in total. In our design, we select a high-performance PMSM product, which is shown in Figure 2.13(a).

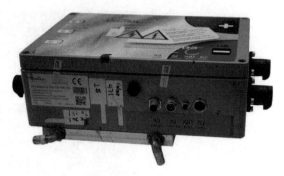

(a) In-wheel Motor   (b) Motor Controller

Figure 2.13: Motor and motor controller.

The maximum torque, speed, and power of the PMSM is 120 Nm, 5,000 rpm, and 60 kW, respectively. Its diameter is 208 mm and its thickness is 80 mm, which is much smaller than products of the same power level. The small dimensions of the PMSM provide possibility for a compact assembly design. In addition, its weight is only 8 kg, which provides the possibility to decrease the unsprung mass to a very desirable level. The water pipe joint of the motor and the joint for the three-phase electric wires are in the inner side of the motor, which provides convenience for the wires configuration and the cooling system design. The motor controller is shown Figure 2.13(b). The weight of the motor controller is 15 kg. There are three kinds of ports on the controller, including the high-voltage three-phase wires ports, the low-voltage control signal wires ports, and the ports for the cooling water. Each in-wheel motor is powered and controlled by its independent motor controller.

The in-wheel motor-driven system is shown in Figure 2.14. Figure 2.14(a) shows the 3D model of the in-wheel motor system. Each component is denoted by the red arrow, including the wheel, wheel center, upright planetary gearbox, brake plate, hub, motor, and inner cover. The wheel size is 18 in. A planetary gearbox is assembled in the wheel to decrease the motor's speed and increase its torque. The motor shown in Figure 2.13(a) is an outer-rotor motor. As an outer-rotor motor, the motor outputs its power by a flange plate. The flange plate is mounted and connected to the sun gear of the planetary gearbox with output shaft by the bolts. The planetary gears are mounted on the hub to drive the wheel. The brake disk is mounted on the hub by stop pins, and the brake caliper is mounted on the upright. Besides the function on mounting the brake disk, the upright is also used to connected to the independent steer and the control arms of the double wishbone suspensions. The upright covers the whole assembly, and the rubbers are mounted between the in-wheel motor assembly and the upright to absorb the vibration and reduce the dynamic load. According to our experience, using rubbers to absorb the dynamic load generated by the road roughness is a practical approach with low cost. Figure 2.14(b) shows the in-wheel motor assembly without upright mounted. In total, there are four in-wheel motor driven systems in the mother vehicle. This design assembles both the motor and the gearbox inside the wheel, which can significantly save the space in the vehicle body. As is widely accepted, this is a great advantage of the in-wheel motor-driven technique.

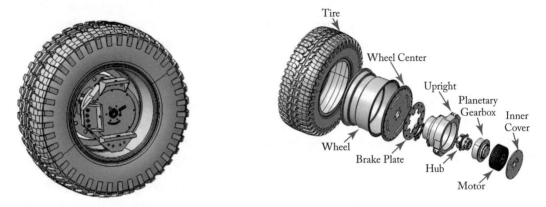

(a) 3D Model of In-wheel Motor System

(b) Assembly without Upright

Figure 2.14: In-wheel motor-driven system.

Before experimenting on the whole UGC, the experiment of the individual in-wheel motor-driven system should be conducted first, since it is a very important subsystem of the UGC. During the experiment of the in-wheel motor-driven system, we should be concerned with two kinds of performance. At first, the power and torque output on the wheel should be tested. Although the motor itself can output 60 kW power, the power and torque on the wheel may be greatly degenerated due to the power loss of the mechanisms. This is also a common approach to evaluate the performance of the planetary gearbox system design. Second, the endurance of the in-wheel motor-driven system should be evaluated. As introduced above, the mechanism in the in-wheel motor-driven system is very easily damaged due to the dynamic load and the rotation of the motor. The best way to evaluate these performance is the dynamometer test.

Figure 2.15 shows the experiments on dynamometer to test the torque and power output of the in-wheel motor-driven system, as well as the reliability and endurance of the system. According to Figure 2.15, the in-wheel motor-driven system is mounted on the dynamometer. The wheel is placed on the drum, and the drum can simulate the resistance force when the vehicle is driving on the road. The human operator controls the laptop to control the motor speed, and the drum can be controlled by the dynamometer's computer to adjust the resistance torque. Therefore, the torque generated by the wheel can be recorded in each speed, and the output power can be calculated. The test results on the dynamometer are shown in Figure 2.16. Figure 2.16(a) shows the torque generated by the wheel, and Figure 2.16(b) shows the power generated by the wheel. The continuous torque value is about 300 Nm, and the peak torque value is 600–700 nm. In addition, it can be seen that the peak torque decreases a little when the motor speed exceeds 4,000 m/min. The peak power of the in-wheel motor-driven system is about 55 kW. As already mentioned, the peak power of the in-wheel motor is 60 kW, which means 5 kW power is lost due to the efficiency of the mechanisms components in the wheel.

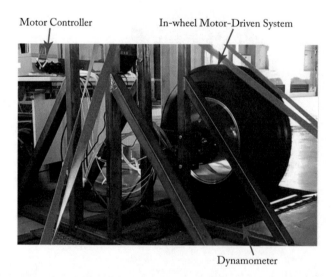

Figure 2.15: The test of in-wheel motor-driven system on dynamometer.

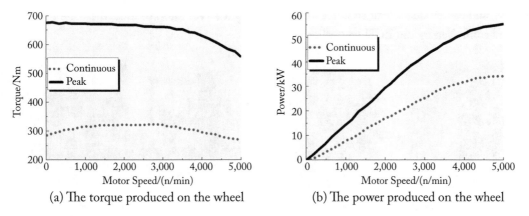

(a) The torque produced on the wheel

(b) The power produced on the wheel

Figure 2.16: The test results of in-wheel motor-driven system on dynamometer.

2.5   DESIGN OF THE INDEPENDENT STEER AND THE DOUBLE WISHBONE SUSPENSION SYSTEMS

2.5.1   DESIGN OF THE INDEPENDENT STEER SYSTEM

The independent steer technique is used in the mother vehicle, which is very important for the four-wheeled UGV to improve the mobility and maneuverability, especially in the narrow spaces in field or urban environments. In the small-sized robots field, the omnidirectional-steer or independent-steer techniques has attracted much research focus. Figure 2.17 shows three kinds of typical omnidirectional steer techniques in the robotics field, including sliding roller wheels, mecanumm wheels, and transformable suspensions [24]. The sliding roller wheel employs a series of slender rollers on the circumference of a wheel that are mounted perpendicular to the rotation direction of the wheel. Therefore, the wheel generates thrust in the wheel's rotation direction and consequently passively slides in the latera direction using the slender rollers. The function of the mecanumm wheels is very similar to the sliding roller wheels. The only difference is that the rollers of the mecanumm wheels are aligned at 45° to the plane of the wheel to generate angular contact forces between the wheels and the ground. The transformable suspension is a more direct way to achieve the omnidirectional steer. It directly adjusts the direction of the suspension control arm to adjust the angle of the wheels to change the moving direction.

However, these techniques cannot be directly applied for heavy-class UGVs, since the load capacity of these mechanisms are relatively small. Therefore, they can only be used for the small sized and light-class mobile robots, and can not be used for the heavy-class UGVs. For the design of the UGC, we use steer linkage and servomotor to achieve the independent steer of each wheel. In other words, each wheel of the mother vehicle of the UGC is steered by an independent steer system, and each independent steer system is driven by a servomotor. In total, there are four servo-motors controlled by the ECU controller.

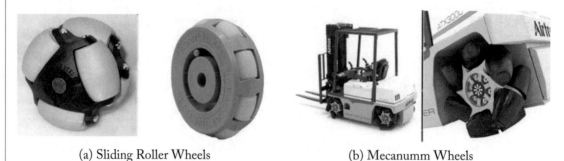

(a) Sliding Roller Wheels   (b) Mecanumm Wheels

(c) Transformable Suspension

Figure 2.17: Typical omnidirectional steer techniques for robots [24].

The main components of the independent steer system are shown in Figure 2.18. The ser-vomotor product is shown in Figure 2.18(a). The maximum power and torque of the servomotor is 0.2 kW and 44 Nm, which is strong enough for the heavy-class UGC. The 3D model and the denotations of the main components of the independent steer system are shown in Figures 2.16(b, c). It should be noticed that, although the maximum torque of the servomotor is relative large, a pinion and rack is still used to further improve the torque of the whole system. The resistance force of each wheel of UGC is relatively large because of the increase of the unsprung mass due to the involvement of in-wheel motor and in-wheel planetary gearbox. In addition, the resistance force between the ground will greatly increase when the UGC is operating in off-road conditions. There-fore, a large steering torque is very necessary to obtain the good steering performance. The details of the pinion and rack system can be seen in Figure 2.16(b).

The servomotor is mounted on the mounting base, which is mounted on the vehicle body frame. A pinon and rack is driven by the servomotor through a flange plate. The pinion is hold in a gear housing, which contains the lubrication oil to lubricate the mechanical system. The rack is mounted in the mounting base, which is welded on the body frame. The wheel is steered by the tierod linkage, which is moved in linear displacement to steer the wheel. One side of the tierod is connected to the rack by joint bearing, and the other side is connected to the upright also by joint bearing. The joint bearing provides enough moving degrees for the whole assembly.

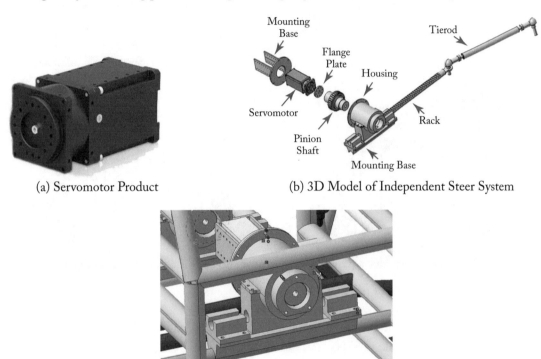

(a) Servomotor Product   (b) 3D Model of Independent Steer System

(c) The Close-up Detail of the Independent Steer System

Figure 2.18: Main components of the independent steer system.

2.5.2   FOUR STEER MODES OF THE MOTHER VEHICLE

Based on the independent steer technique, each wheel of the mother vehicle can be controlled independently. The mother vehicle of the UGC can achieve different steer modes to negotiate different narrow environments. In practice, the UGC has four steer modes, including Ackermann Steer Mode (ASM), Double Axle Steer Mode (DASM), Diagonal Move Steer Mode (DMSM), and Zero Radius Steer Mode (ZRSM), shown in Figure 2.19. Each steer mode can be used in different conditions, which will be detailed in the following sections.

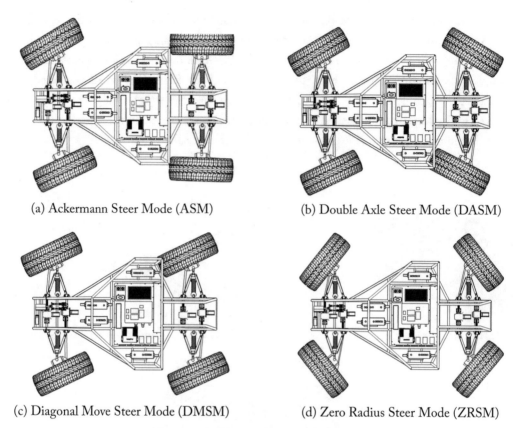

(a) Ackermann Steer Mode (ASM)   (b) Double Axle Steer Mode (DASM)

(c) Diagonal Move Steer Mode (DMSM)   (d) Zero Radius Steer Mode (ZRSM)

Figure 2.19: Four steer modes of the UGC.

The ASM is very commonly used in the passenger car field. In ASM, only the front wheels are steered to yaw the vehicle as Figure 2.19(a) shows. For the UGC, the ASM should be used in high-speed conditions to improve the handling stability. If other steer modes are used in high-speed conditions, such as the DASM or DMSM, the vehicle may be easy to lose the stability and spin. In DASM, both the front and rear wheels are steered in opposite directions, which can significantly reduce the steer radius. Therefore, the DASM should be used for low-speed sharp turning condition to improve the maneuverability in narrow space. It is widely known that it is very important for the military vehicle to have great mobility to operate in narrow space. In DMSM, both the front and rear wheels are steered in the same directions and same angles. Therefore, the vehicle can move diagonally or laterally without yaw motion to improve its maneuverability. This maneuver is also very important for the military vehicle in some special missions. In ZRSM, which is a unique advantage produced by the four-wheel independent steered technique, the left and right wheels can steered in opposite directions, as Figure 2.19(d) shows. The position of the wheels can be placed to a predefined positions to make the turning center locates at the middle of the vehicle to avoid tire skidding. The detail of the control strategy in each steer mode will be described in Chapter 3.

(a) ASM in Civilian and Military Version

(b) DASM in Civilian and Military Version

(c) DMSM in Civilian and Military Version

(c) ZRSM in Civilian and Military Version

Figure 2.20: Pictures of the four steer modes of the UGC.

The pictures of the four steer modes of the UGC are shown in Figure 2.20. These pictures are taken with the civilian version and military version, respectively. The significance of the ZRSM should be further discussed. Actually, ZRSM is an important advantage achieved by the independent steer technique. In ZRSM, all the wheels are placed in the positions as Figure 2.20(d) shows, which enables the UGC to achieve pivot steer to greatly improve the mobility in narrow spaces. As introduced above, the six-wheeled skid steered vehicle, such as MULE and Crusher, can also achieve pivot steer. However, there are apparent drawbacks for the six-wheeled skid-steered pivot steer. At first, the skid-steered pivot steer technique uses the torque and speed difference between each side wheels to steer the vehicle, and the wheels will not be steered during the pivot steer. Therefore, large tire slip and skid is unavoidable during the pivot steer. It may cause the difficulty to control the trace and position of the vehicle. This causes large motion uncertainty for the skid-steered vehicle. Second, the vehicle's tires will be badly worn and the tires' lifetime will be greatly reduced due to the large tire slip and skid, which causes difficulty and large cost for the maintenance. As is widely known, the convenience for maintenance is very important for the military vehicle.

However, the ZRSM achieved by independent steer technique can successfully solve these problems. As shown above, the wheels can be placed in the predefined positions to make the pivot turning center at the middle of the vehicle. When the human operator controls the UGC to pivot steer, if the left-hand turn is selected, the motor torque of left-side wheels will be negative and the motor torque of right-side wheels will be positive. The speed of the pivot steer can be adjusted by the position of the lever on the command station, which represents the amount of the input motor torque value. In ZRSM, the vehicle can pivot steer smoothly and the tire slip and skid can be significantly reduced because the wheels have been steered. Therefore, the handling performance can be greatly improved to easily control the vehicle position and maneuver time during the pivot steer, and also the problem of the tire wearing can be successfully solved.

2.5.3   DESIGN OF THE DOUBLE WISHBONE SUSPENSION

As a heavy-class military UGV, it's necessary to select a mature suspension type of ground vehicles. As introduced above, although many advanced and novel suspension has been proposed in mobile robot field, they can not be used for heavy-class UGV due to the limit of the load capability. The double wishbone type suspension is selected for the UGC. As widely accepted, the independent double wishbone suspension can significantly improve the handling performance of the vehicle, especially in off-road conditions. The control arms of the double wishbone suspension have a better ability than those of other types of suspension to control the trace of the wheel to achieve the desirable kinematic relationship, such as the variation of camber angle, toe angle, caster angle, and kingpin inclination angle during the vertical wheel movement. Therefore, the double wishbone type is selected during the design, since the handling performance is very important for a military vehicle. In addition, since the steering system is not manipulated by the human, the aligning moment of the steering system can be neglected during the design. Therefore, the kingpin inclination and caster angle is designed as zero to reduce the manufacturing difficulty of the upright and other components. The camber gain and toe angle gain of the suspension is design to be negative. Figure 2.21(a) shows the overview of the suspension. Four air shock absorbers, which have 500 mm vertical travel distance, are used to improve the performance in off-road conditions. Figures 2.21(b, c) shows the 3D model and picture of 1/4 suspension. The upper and lower arms are connected to the upright by the sphere joints. As the above section introduced, the in-wheel motor-driven system is assembled in the upright. The cooling water pipes and electric wires are connected to the motor inside the wheel. The water pipes and electric wires have to go through the suspension from the body to the wheel. In practice, the water pipes and the electric wires are covered by the protect materials to avoid the damage. In Figure 2.21(c), the protect material is removed to clearly show the components.

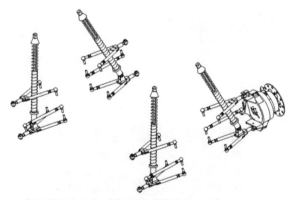

(a) 3D Model of Double Wishbone Suspension

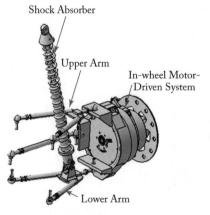

(b) 3D Model of 1/4 Suspension

(c) Picture of Suspension

Figure 2.21: Double wishbone suspension.

2.6   DESIGN OF THE MECHANISMS FOR CARRIER-BASED ROBOTS OR ROTORCRAFTS

2.6.1   DESIGN OF THE MECHANISMS FOR SENDING OUT OR TAKING BACK THE ROBOT

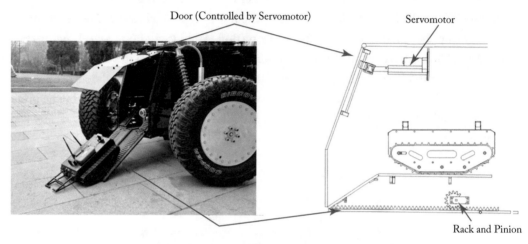

(a) Side View

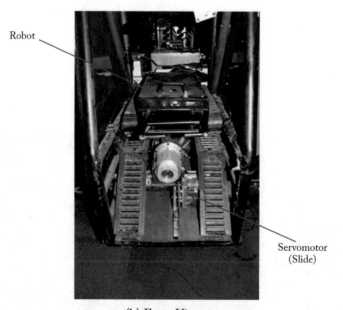

(b) Front View

Figure 2.22: The mechanism to send out or take back the robot.

As a prototype model, the purpose of the UGC is to demonstrate the concept of the UGC. Therefore, both the military and civilian version UGC don't carry too much carrier-based equipment. As introduced in the section above, in the military version UGC, only an armed rotorcraft, a twin duct aircraft, and a loitering munition are carried on the UGC. In the civilian version UGC, only six rotorcrafts and a tracked robot are carried on the UGC. When the UGC is applyied in real applications, the carrier-based equipment may be adjusted according to the real requirements, such as the adjustment of the amount of the rotorcrafts.

At first, the mechanism to send out and take back the robot is important, which should enable the robot to go out and go in the body quickly and easily enough. The mechanism to send out and take back the robot is shown in Figure 2.22. The key components of the mechanism are a door, servomotors, and a slide. When the robot is on board, it is located at the rear side of the mother vehicle. The door at the front of the vehicle is closed. The robot is stabilized in the vehicle body by magnets. If the robot is not stabilized by the magnets, it will be moved when the UGC is working on the uneven robot. When the human operator determines to send out the robot, the door can be controlled to be open by a servomotor, and the switch is on the command station, which can be handled by the human operator. A slide can be controlled to be sent out to help the robot to get off the mother vehicle. As Figure 2.22 shows, the door is controlled to be open as about 90°. The slide is putted out and contacted to the ground, which makes it easier for the robot to be sent out. The slide is controlled by the rack and pinion. The robot itself is not designed and manufactured by the authors. A robot product is selected and modified to add communications and camera systems. The pictures of the robot product can be seen in the Figure 2.23.

Some examples are shown in Figure 2.20 to illustrate how the robot works, such as the examples in the building and field environments. A radio station and a camera are carried on the robot to provide the communication and image feedback with the mother vehicle and the command station.

(a) Robot Climb on the Stairs

(b) Camera Feedback

(c) Robot in the Field

(d) Camera Feedback

(e) Robot in the Building

(f) Camera Feedback

Figure 2.23: Illustration of how a robot works.

2.6.2   DESIGN OF THE MECHANISMS FOR LANDING AND STABILIZING THE ROTORCRAFTS

The platform, on which the rotorcrafts are landed, is shown in Figure 2.24, which uses the civilian version as the example to show the design. Actually, there are many approaches that can be used to grab and stabilize the rotorcrafts, such as the grab mechanisms. However, the grab mechanisms are too complex and expensive. Therefore, a cheaper and practical approach is used for the UGC: the electromagnets. The platform is mounted on the top of the body frame by the bolts, and several electromagnets are mounted inside the platform. The material of the bracket of the rotorcrafts is modified to iron. When the rotorcrafts are on board, they can be stabilized by the electromagnets. When the human operator determines to send out the rotorcrafts, the electromagnets can be controlled to be turned off by switches on the command station, and the rotorcrafts can take off. In addition, the control levers of the robot and the rotorcrafts are also assembled on the command station, which provides the possibility for the human operator to control the mother vehicle, rotorcrafts, and robot, respectively. Figure 2.25 shows the illustration of the rotorcrafts to take off or land on the platform.

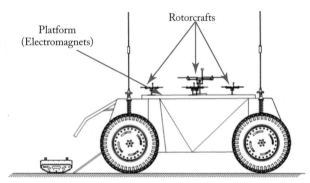

Figure 2.24: The platform for the rotorcrafts.

Figure 2.25: Illustration of the rotorcrafts at take off or landing.

CHAPTER 3

Advanced Chassis Dynamics Control of the Unmanned Ground Carrier

In the previous chapter, the overall design of the UGC is described, as well as the design of the important subsystems, such as the high-voltage battery system, in-wheel motor-driven system, and independent steer system. For a full x-by-wire vehicle, the control system plays the most important role to improve the overall performance of the vehicle. Therefore, in this chapter, the chassis dynamic control of the UGV will be described, including the overall control architecture block and the detail control strategy of each subsystem.

The research of the vehicle dynamics control is always a popular topic in the academic field. With the rapid development of vehicle electronics technique, numerous vehicle dynamics control techniques are emerging, such as Traction Control System (TCS), Antilock Brake System (ABS), Electronic Stability Program (ESP), Active Front-steer System (AFS), Rear Wheel Steer (RWS), or Direct Yaw Control (DYC) [25–32]. In recent years, with the development of drive-by-wire technique, Four Wheel Independently Actuated (FWIA) EV has become a most widely focused configuration with unique features. For FWIA EV, the motor torque acting on each wheel can be controlled independently with quick response, which enables the designers to achieve advanced integrated dynamics control [33, 34]. Therefore, FWIA EV has become a desirable platform to apply and demonstrate various individual or integrated dynamics control systems through x-by-wire technique. The literatures about dynamics control of FWIA EV can provide a clear and comprehensive observation of the state of art in the academic field.

As a basic advantage of FWIA EV, each wheel of the FWIA EV is drive-by-wire and driven by an independent motor. Therefore, each wheel's torque can be independently and precisely controlled. It provides great convenience for achieving TCS or optimal tire slip ratio control. Many control approaches have been applied, such as optimal control, fuzzy control, switch control, robust control, or adaptive control [35–38]. The UOT March II, which has been mentioned in Figure 2.1, was the first famous FWIA EV to demonstrate the traction control technique. The readers can find the details in the literature. In addition, the lateral dynamics control is always an important topic among the vehicle dynamics control techniques. For the lateral stability control of FWIA EV, DYC is one of the most effective techniques, which utilizes the independent motors to provide additional active yaw moment to improve the handling stability in critical cornering condition. Shibahata first proposed the basic principle of DYC system [39]. After that, Nagai proposed a MMC controller to control the vehicle to follow the desired 2 Degree of Freedoms (2-DoFs) dynamic model with the feedback of yaw rate and side slip angle [40]. Hori proposed a robust MMC controller. He further claimed that the research should focus on how to calculate the external yaw moment to make the vehicle follow the desired vehicle model [41]. Like Hori stated, the following researchers mainly placed their focus on the control method to calculate the yaw moment. To track the desired vehicle lateral dynamics response calculated by the 2-DoFs vehicle dynamic model, many control approaches have been proposed to calculate the desired yaw moment, including optimal control, sliding mode control, adaptive PID control, and H-infinity control [42–46]. Moreover, with the feedback of independent motors information, such as motor torque or rotation speed, several DYC control blocks with vehicle parameters online estimation have been proposed to improve the control performance in extreme condition, especially the DYC blocks with tire cornering stiffness estimation, tire-road friction coefficient estimation, body side slip angle estimation, or tire lateral force estimation [47–50]. As a unique problem of FWIA EV, the control allocation of traction force between individual wheel and fault-tolerant control has been widely investigated. Many torque distribution approaches have been proposed aims at simultaneously assuring the handling stability and maximizing the tire adhesion capability, or improving energy efficiency while taking power losses of independent motors as cost function. The reallocation approaches of traction force when motor failure occurs are investigated with the consideration of tire-road friction change or other uncertainties [50–54]. To deal with the parametric uncertainties and time-varying problems in DYC system, the controller design with consideration of tire cornering stiffness uncertainties, time-varying delay of CAN network or data dropout of sensors are investigated. Several robust controllers are proposed based on norm bound, Linear Matrix Inequalities (LMI), or H2/H-infinity approaches [55–59].

On the other hand, with the development of the steer-by-wire technique, the 4WS technique has also been widely focused. Based on the 4WS technique, the decoupling and independent control of the yaw rate and sideslip angle provides the possibility for significantly improving the handling, comfort, and lane-keeping performance of the vehicle [60–62]. With lateral speed observation or directly measuring, several control strategies have been proposed to decouple the lateral and yaw dynamics of the vehicle, including PI control, nonlinear decoupling control, robust control, H∞ control, and hybrid control [63–66]. In recent years, the integrated dynamics control with both steer-by-wire and drive-by-wire techniques has attracted increasing research focus. The combination of 4WS and AWID can significantly increase the additional yaw moment range to regulate the vehicle lateral motion, which is proved to effectively improve the agility, maneuverability, and handling stability of the vehicle [67–69]. Generally speaking, an integrated 4WS/4WD controller is functionally divided into two parts: Motion Control (MC) and Control Allocation (CA) parts. The MC aims at calculating the desired vehicle response, which can be defined by desired yaw rate, sideslip angle, or a combination of them. As a highly overactuated system whose four-wheel torques and steering angles can be independently controlled, the CA algorithm has redundant 2-DoFs to achieve extra performance, such as minimization of the tire workloads or the energy consumption. Based on this, many researchers deal with the problems of physical saturation of actuators, time-varying delay of network, data dropout of sensors, tire cornering stiffness uncertainties, or other parametric uncertainties, and many control approaches have been proposed, such as Linear Quadratic Regulator (LQR), Lyapunov, energy-to-peak, Linear Matrix Inequalities (LMI), gain scheduling, and H2/H-infinity approaches [70–76].

3.1   THE APPLICATION OF CONTROL CONFIGURED VEHICLE (CCV) CONCEPT

3.1.1   BASIC CONCEPT OF CCV

During the overall configuration design and the control system design of the UGC, the CCV concept is applied to improve the overall configuration flexibility of the vehicle, which will be discussed in this section.

Actually, the CCV concept has been widely focused and has been applied in the field of aircraft to integrate the control and the mechanical systems, which aims at forming a new overall configuration concept and greatly improving the aircrafts' performance through the X-by-wire flight control techniques [77, 78]. The active flight control techniques of the aircrafts under CCV can be divided into relaxed static stability, maneuver load control, and envelope limit control, etc. Currently, the application of CCV has been an important feature of the advanced aircrafts, especially the military fighter aircraft.

Before the emergence and application of the CCV, the design process of the control system and mechanical system of aircrafts were independent of one another. In other words, after the final mechanical hardware configuration is selected and determined, then various flight control systems are added, which have no effect on overall mechanical hardware configuration. Consequently, the mechanical configuration flexibility and the overall performance is limited. To the authors' knowledge, it has to be accepted that the field of ground vehicle dynamics control is in the same period of aircraft design before the emergence and application of CCV. For the design of ground vehicles, the designers primarily only consider the mechanical hardware subsystems, such as vehicle body, engine, transmission, and chassis subsystems. After the final mechanical hardware configuration is obtained and determined, a variety of vehicle dynamics control techniques are applied independently.

The research on the vehicle dynamics control was summarized briefly above. However, the research in the aforementioned works still only concentrated on the design of control strategy and algorithm of vehicle dynamics control system. In other words, the process of the design of the ground vehicle's mechanical hardware configuration and the vehicle control system are independent of one another just like the period of aircrafts before the application of the CCV. Therefore, the potential to improve hardware configuration flexibility and the performance of the ground vehicle is

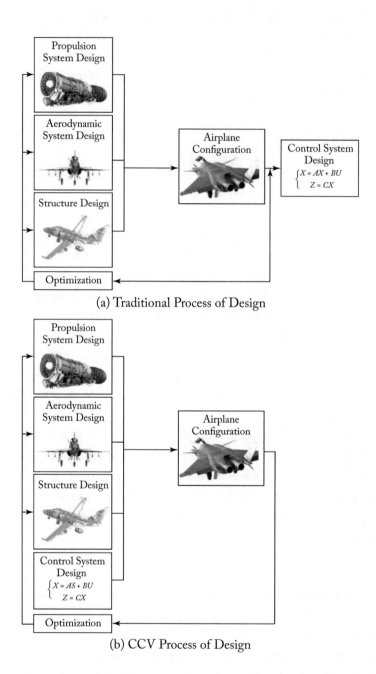

(a) Traditional Process of Design

(b) CCV Process of Design

Figure 3.1: Comparison of aircraft design process.

consequently limited. Actually, with the development of the UGVs, which mainly apply x-by-wire techniques, the requirement of hardware configuration flexibility is rapidly increasing since there can be no human in the vehicle and much more electric subsystems are involved because of the application of steer-by-wire or drive-by-wire techniques. To this end, the CCV concept is applied during the overall design of the UGC, which aims at forming a novel configuration principle to improve the configuration flexibility. The details of the application of CCV can also be seen in our previous work [79].

At first, how CCV works in aircraft field should be introduced and discussed. Based on the active flight control and X-by-wire techniques, CCV integrates the control and mechanical systems to achieve the overall hardware configuration selection approach to improve the overall performance. Figure 3.1 shows the design process under traditional approach and CCV concept of aircraft. In the process of traditional design, it can be seen that the design of control system is on the outside of the hardware configuration loop. The control system design has no effect on the hardware configuration. The control system is not determined until after the final hardware configuration has been determined and selected. Figure 3.1(b) shows the design process of CCV. It involves the control system in parallel with other mechanical systems for final hardware config-uration. Therefore, the design of the control system is able to directly affect the hardware configu-ration. The integration of control and mechanical systems makes contribution to the maximization of the aircraft's overall performance.

Take relaxed static stability (RSS) for example, which is a very typical dynamics control strat-egy of CCV in aircraft field. For traditional aircraft, the center of the aerodynamic force has to be in the back of the center of gravity (C.G.) location to obtain enough inherent static stability, which causes great down tail force loads to make enough pitch moment balance. By applying CCV, if a feedback dynamics control system is applied to provide enough artificial stability, the longitudinal stability of the aircraft can be relaxed. Consequently, the C.G location can be placed in the back of the location of aerodynamic center, which can significantly improve the hardware configuration flexibility. The down loads tail can become an up-direction loaded tail to greatly reduce the wing loads, and also the size of tail can be reduced to reduce fuel consumption.

Currently, the design process of ground vehicles is the same as the traditional design pro-cess of aircraft in Figure 3.1(a). The vehicle designers mainly consider the mechanical systems during the hardware configuration selection of the ground vehicle, such as the propulsion and the transmission systems, the suspension and the steering systems, the body system, etc. The design of the control system is independent of the design of the mechanical systems. After the final vehicle overall hardware configuration is determined, then the control systems are added independently.

Consequently, the potential to improve the hardware configuration flexibility and the overall performance is limited. As x-by-wire UGVs emerge, the requirement of hardware configuration flexibility is much greater than ever before because there can be no human being in it and more electric systems are involved because of the application of the x-by-wire technique. The traditional design process of the vehicle cannot satisfy the requirement of these vehicles. Therefore, it is very necessary to apply CCV in the design of the x-by-wire ground vehicle. The control system of the vehicle should be designed and selected in parallel with mechanical hardware systems for the final configuration to improve the hardware configuration flexibility and the overall performance.

The difference between the traditional and CCV design process for the ground vehicle should be further emphasized. The traditional design process only considers the mechanical hardware systems. Therefore, the vehicle dynamics system, which is constructed by the propulsion, transmission, suspension, steer, and body systems, should be designed inherent stable. Take the lateral dynamics, which affect the handling performance of the vehicle, as an example. In traditional vehicle dynamics theory, the lateral dynamics system should be understeer (inherent stable), which can be described by Static Margin (S.M.):

$$\text{S.M.} = \frac{c_r}{c_f + c_r} - \frac{l_f}{L} , \tag{3.1}$$

where c_f and c_r is front and rear tire cornering stiffness, respectively. l_f is distance from front axle to C.G. L is wheel base.

When S.M. is higher than 0, the vehicle is understeer. The higher the S.M. , the higher static stability the vehicle, and the more understeer the vehicle is. The S.M. value depends on the mechanical configuration specifications, especially the C.G location. Consequently, the structure configuration flexibility will be strictly limited due to the positive requirement of S.M. However, if CCV concept is applied, this situation will be changed. If the control system is considered in parallel with the mechanical systems, the dynamics system of the mechanical systems won't need to be designed inherent stable. It can be designed inherently unstable, which means the S.M. can be designed negatively to improve the configuration flexibility. In addition, the whole system, which consists of both mechanical and control systems, can be designed as closed-looped stable to improve the handling performance.

3.1.2   THE HARDWARE CONFIGURATION PROBLEM OF THE UGC WITHOUT CCV

The application of full x-by-wire technique poses a big challenge for the overall configuration of the UGC. Compared to a traditional vehicle, many new components are involved due to the application of the x-by-wire techniques. For example, due to the application of electric power technique, a 160 kg battery pack is involved. Due to the the application of the four-wheel independent-drive technique, four 80 kg in-wheel motor driven systems are involved, as well as four 15 kg motor controllers. Due to the application of independent steer, four 4 kg servomotors are involved. Due to the application of collaborative control technique, a 110 kg platform, six 3 kg rotorcrafts, and a 20 kg robot are involved. These new systems and components cause challenge for the overall configuration.

The biggest problem is the selection of the C.G location. According to traditional understeer requirement as Equation (3.1) shows, the C.G location should be designed in the front of the vehicle to improve the understeer characteristics. Actually, in traditional vehicles it is easy to achieve understeer, since the engine and transmission systems are usually located at the front of the vehicle, which decreases the value of l_f, and consequently improves the understeer characteristics. However, for the new configured X-by-wire vehicle, such as UGC, it is very difficult to achieve understeer. For example, the heavy battery pack should be placed in the middle of the vehicle to provide convenience for other powered instruments, which may cause a back C.G location and degenerate the understeer.

The main specifications of the UGC were shown in Chapter 2. By simple calculation, it can be known that when S.M. value equals 0, l_f and l_r should be 1 m and 1.2 m. In other words, if traditional understeer requirement is applied, the lf value should be smaller than 1 m and the lr value should be larger than 1.2 m. The understeer requirement strictly limits the C.G location and further limits the overall configuration.

Figure 3.2 illustrates the limit of the overall configuration under understeer requirement. Given the wheelbase L and track width B, the positions of in-wheel motor driven systems, double wishbone suspensions, and independent steer motors are almost fixed. Therefore, only the locations of battery pack, platform for rotorcrafts, and motor controllers should be selected. According to positive S.M. requirement ($l_f \leq 1$ m, $l_r \geq 1.2$ m), after calculated, Figure 3.2 can be drawn to illustrate the situation of the configuration when $l_f = 1$ m and $l_r = 1.2$ m.

In Figure 3.2, the body frame is not drawn to clearly show the configuration of the components. The battery pack should be placed at the front to obtain front C.G location. The allowable range for the battery pack according to understeer requirement is denoted by a red box. The battery pack should be allowed to be moved forward to further reduce the l_f. However, the allowable range is only 11 cm. If it is moved forward more than 11 cm, it will contact to the steer servomotor. The position of servomotor cannot be changed, since it should be placed in an appropriate position to make the tierod straight. In addition, not only the battery pack, but also the positions of low voltage instruments and sensors, which are mounted on the top of battery pack, will be limited. However, an important advantage of the x-by-wire technique is that the designer can place the components to anywhere as he wants. The understeer requirement limits this advantage.

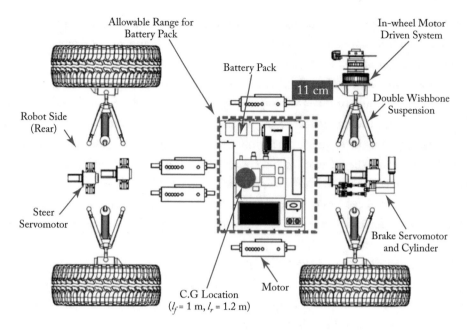

Figure 3.2: Illustration of overall configuration under understeer requirement.

3.1.3   THE HARDWARE CONFIGURATION OF THE UGC UNDER CCV

According to Figure 3.3, if CCV is applied for the UGC, the overall configuration flexibility could be greatly improved. The CCV allows the vehicle lateral dynamics system to be inherently unstable. Therefore, the S.M. value can be designed negative. Consequently, the battery pack and the platform of the UGC can be placed in any position the designer wants. The x-by-wire technique eliminates the mechanical links in the vehicle. Therefore, all the components should be flexibly placed to any position, where can improve the space saving, wires configuration, and assembly convenience, or satisfy other requirements. Apparently, if CCV is not applied, this advantage will be limited.

Under CCV concept, we can select the positions for the components according to various practical concerns without the positive S.M. requirement in mind. The detail of design process is neglected, and the final overall configuration was given in Chapter 2. The major difference between the final configuration and the understeer configuration is: (1) the battery pack is placed in the middle back of the vehicle; and (2) the platform for rotorcrafts is placed in the middle top of the vehicle. The advantage of the final configuration is: (1) the charging ports and the BMS monitor of the battery pack are placed in the middle of the vehicle (right side). Therefore, it is convenient to charge the battery pack and read the data of the BMS; (2) the platform is mounted in the middle of the vehicle (on the top), which provides convenience for the rotorcrafts landing and improves the shape looking; and (3) the battery pack is closer to the robot, which is located in the rear side of the vehicle, to provide convenient charging of the robot.

After the final overall configuration is determined, the final C.G location can be calculated. The l_f value is 1.4 m and l_r value is 0.8 m. The S.M. value can be calculated as -0.1818, which is inherently unstable (oversteer). It is unacceptable according to traditional understeer requirement, however, it is acceptable under CCV. Although the mechanical systems are designed inherent unstable to improve configuration flexibility, the control system should be involved to assure the closed-looped stability. This principle can be illustrated by following equations. Consider the 2-DoFs vehicle lateral dynamics model for traditional vehicle without any control system involved [79].

$$\dot{x}(t) = Ax(t) + B_1\delta(t), \tag{3.2}$$

where

$$x(t) = \begin{bmatrix} \beta \\ r \end{bmatrix}, \quad A = \begin{bmatrix} \dfrac{c_f + c_r}{mu} & \dfrac{l_f c_f - l_r c_r}{mu^2} - 1 \\ \dfrac{l_f c_f - l_r c_r}{I_z} & \dfrac{l_f^2 c_f + l_r^2 c_r}{I_z u} \end{bmatrix}, \quad B_1 = \begin{bmatrix} -\dfrac{c_f}{mu} \\ -\dfrac{l_f c_f}{I_z} \end{bmatrix}, \tag{3.3}$$

where β is the sideslip angle, r is the yaw velocity, δ is the steering angle, l_r is the distance from rear axle to C.G, m is the vehicle mass, I_z is the yaw inertia, and u is the vehicle speed.

For the UGC, several techniques can be used to provide additional yaw moment to regulate the lateral dynamics behavior of the vehicle, such as the yaw moment provided by the in-wheel motors or independent steer techniques. Consider the 2-DoFs dynamic model with additional yaw moment:

$$\dot{x}(t) = Ax(t) + B_1\delta(t) + B_2 M(t), \tag{3.4}$$

where $M(t)$ is the additional active yaw moment. $B_2 = [0\ 1/I_z]^\mathrm{T}$. The feedback control law can be constructed as $M(t) = Kx(t)$, and assume $K = [K_1, K_2]$. After manipulation, the characteristics matrix of the closed-looped system can be obtained as:

$$A = \begin{bmatrix} \dfrac{c_f + c_r}{mu} & \dfrac{l_f c_f - l_r c_r}{mu^2} - 1 \\ \dfrac{l_f c_f - l_r c_r}{I_z} + \dfrac{K_1}{I_z} & \dfrac{l_f^2 c_f + l_r^2 c_r}{I_z u} + \dfrac{K_2}{I_z} \end{bmatrix}. \tag{3.5}$$

According to the characteristics matrix of the original vehicle dynamics system (matrix A) and the characteristics matrix of the control-involved vehicle dynamics system (matrix A_0), the CCV concept can be further understood. For the original vehicle dynamics system, the stability is only determined by the mechanical specifications m, L, I_z, C_f, C_r, l_f, l_r, which leads to the understeer requirement. However, for the control-involved vehicle dynamics system, the stability is not only determined by the mechanical specifications, but also by the control parameters K_1 and K_2. Therefore, the mechanical systems can be designed inherent unstable to improve the flexibility, and the control system can be used to assure the closed-looped stability. If CCV is not applied and the control system is not involved, the UGC will be easy to lose the stability due to the inherent unstable characteristics. As widely known, there is unstable limit speed existing for the inherent oversteer vehicle. According to the specifications of the UGC, it can be calculated that the limit speed of the UGC is 67 km/h. In other words, if no control system is involved, the UGC will lose the handling stability in the maneuver higher than 67 km/h speed.

In conclusion, the benefit of the CCV can be clearly shown in Figure 3.3. The C.G location when S.M. equals 0 is illustrated by the red circle. When the C.G location moves forward, the vehicle will be understeer. When it moves backward, the vehicle will be oversteer. As Figure 3.3 illustrates, the understeer requirement limits the C.G location in the green box. However, if CCV concept is applied, the lateral dynamics system is allowed to be designed as inherent oversteer, so that the C.G location can be placed in both blue and green boxes. The blue circle indicates the C.G location of the final application. Therefore, the application of CCV increases the flexibility of the C.G location a lot (l_f from 1 m to 2.2 m).

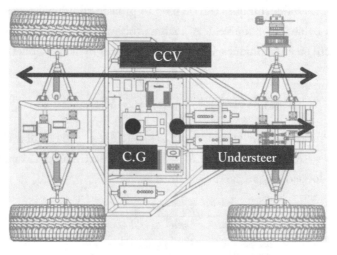

Figure 3.3: Acceptable C.G range under understeer requirement or CCV.

3.2   OVERALL CHASSIS DYNAMICS CONTROL ARCHITECTURE

(a) Human operator controlling the UGC through the command station

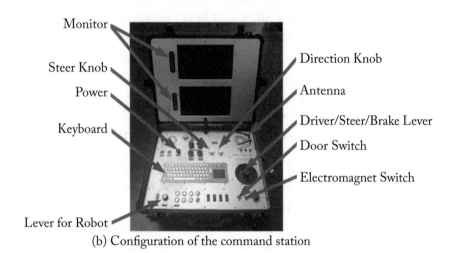

(b) Configuration of the command station

Figure 3.4: Remote control of the UGC.

Generally, the UGV can be operated in both autonomous and remote control modes, which will be determined according to different missions' requirements. This was discussed in the Chapter 1. However, the UGC is mainly used in remote control mode according to the view of the practice. The reason should be discussed here. As mentioned above, the UGC is mainly used in military applications. In military applications, the working scenes are much more complex than that of civilian applications. The complex environment makes the environment perception very difficult. Therefore, the autonomous mode is not practical for a military-used UGV currently. Based on this reason, the UGC is mainly used in remote control mode to satisfy the requirement of the practice.

Figure 3.4(a) shows the human operator controlling the UGC through the command station. Based on telecommunication, remote control, and image transmission techniques, the UGC can be controlled by the command station to conduct the missions. Figure 3.4(b) shows the configuration of the command station. A monitor is assembled on the command station with the image feedback by the cameras, as well as some functional knobs, levers, and switches. In this book, the remote control strategy of the UGC will be detailed in the view of practice.

The hierarchical dynamics control block in remote control mode is shown in Figure 3.5. The first layer is the signal from the human operator. The second layer is the command station. The knobs, switches, and levers on the command station was previously detailed. The most commonly used knobs, switches, and levers on the station are shown in Figure 3.5, including the selection knob for steer mode and positive direction, switches for the door and electromagnet, levers for steer, drive and brake. In addition, there are also the control levers for the robot and rotorcraft. According to Figure 3.5, the upper controller receives the control signal from the command station and calculates the target angle of steer servomotor, target traction force, and desired yaw moment. If the ZRSM is selected, the wheels will be automatically placed in the predefined ZRSM positions. In ASM, DASM, and DMSM, the position of the steer lever is in linear proportion to the target angle of the servomotors. The relationship between the steer servomotor angle and steer lever is:

$$\begin{cases} P_{s,ASM} = K_1\delta_{11} = K_1\delta_{12} \\ P_{s,DASM} = K_1\delta_{11} = K_1\delta_{12} = -K_1\delta_{21} = -K_1\delta_{22} \\ P_{s,DMSM} = K_1\delta_{11} = K_1\delta_{12} = K_1\delta_{21} = K_1\delta_{22}, \end{cases} \tag{3.6}$$

where $P_{s,ASM}$, $P_{s,DASM}$, and $P_{s,DMSM}$ is the percentage of the steer lever in ASM, DASM, and DMSM. δ_{11}, δ_{12}, δ_{21}, and δ_{22} is the angle of the steer servomotor for front left, front right, rear left, and rear right wheel, respectively. K_1 is the control gain.

The value of the target total traction force is in proportion to the position of the drive lever. The active yaw moment provided by the in-wheel motors is applied to improve the performance of the mother vehicle, which will be detailed later. In ZRSM, the traction force of left and right sides wheels is in opposite directions, which is determined by the turn direction. The upper controller outputs the value of target traction force and desired yaw moment. In lower controller, a force distribution algorithm is adopted to calculate the target longitudinal force of each wheel. A slip ratio controller is used to adjust the motor torque to negotiate complex terrain. The vehicle state feedback includes vehicle speed, yaw velocity, sideslip angle, motor torque, speed, and damper displacement. The algorithms to calculate the active yaw moment and distribute the traction force of each motor are the most important parts. As Figure 3.6 shows, the traction force acting on wheel is divided into two parts, which are determined by the target traction force value and the desired yaw moment value, respectively.

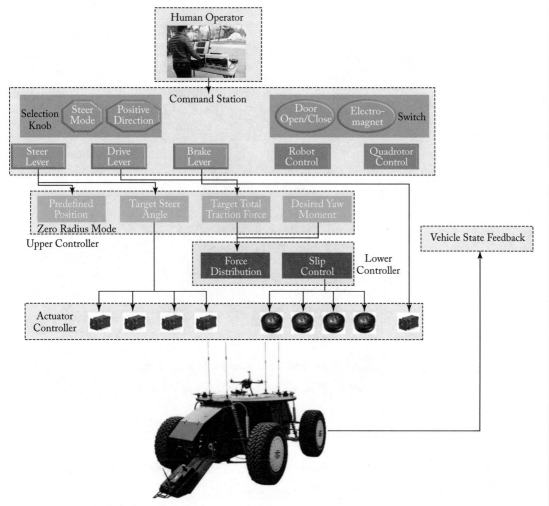

Figure 3.5: Hierarchical chassis dynamics control block.

$$F_{xij} = F_{xTij} + F_{xMij}, \tag{3.7}$$

where F_{xij} (i,j = 1,2) are the actual longitudinal force acting on four wheels, respectively; F_{xTij} (i,j = 1,2) are the part of the longitudinal force determined by the target total traction force; and F_{xMij} (i,j = 1,2) are the part of the longitudinal force determined by the desired yaw moment.

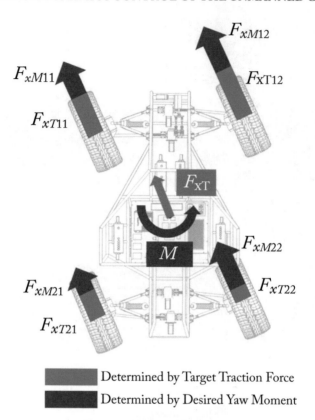

Figure 3.6: Illustration of tire traction force distribution.

The yaw moment controller, the traction force distribution algorithm, and the tire slip ratio controller will be shown in the next few sections.

3.3   CONTROL-ORIENTED UNCERTAIN VEHICLE LATERAL DYNAMICS MODEL

3.3.1   BASIC 2-DOFS VEHICLE LATERAL DYNAMICS MODEL

Figure 3.7 shows the illustration of the vehicle dynamics model. It is assumed that the vehicle is moving on a surface with a local coordinate frame denoted by (O, X, Y) assigned at its C.G. u, v represents the longitudinal and lateral vehicle speed. β, r is the side slip angle and the yaw velocity. δ_{11}, δ_{12}, δ_{21}, δ_{22} represents the steering angle of four wheels. F_{y11}, F_{y12}, F_{y21}, F_{y22} represents the lateral force of four tires. α_{11}, α_{12}, α_{21}, α_{22} represents the lateral slip angle of four tires.

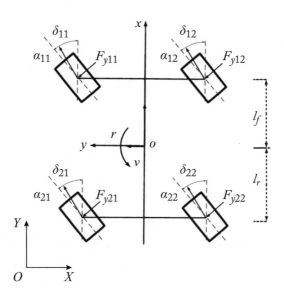

Figure 3.7: Vehicle lateral dynamics model.

In this chapter, the 2-DoFs vehicle lateral dynamics model of the UGC will be used for controller design. To obtain the 2-DoFs model, the following assumptions are made. (1) Assume F_{yf} is the sum of the lateral force of front tires, and F_{yr} is the sum of the lateral force of rear tires. (2) Assume δ_f, δ_r is the steering angle of front and rear wheels. (3) Assume α_f and α_r is the tire slip angles of front and rear tires. (4) Assume M is the active yaw moment provided by the independent motors. The single-track 2-DoFs vehicle dynamics model of the UGC can be obtained:

$$\begin{cases} mu\dot{\beta} = F_{yf} + F_{yr} - mur \\ I_z\dot{r} = l_f F_{yf} - l_r F_{yr} + M \ . \end{cases} \tag{3.8}$$

To obtain linear 2-DoFs model, the tire lateral force can be described based on linear assumption:

$$\begin{cases} F_{yf} = c_f \alpha_f \\ F_{yr} = c_r \alpha_r \ . \end{cases} \tag{3.9}$$

The α_f and α_r can be described as:

$$\begin{cases} \alpha_f = \beta + \dfrac{l_f r}{u} - \delta_f \\ \alpha_r = \beta + \dfrac{l_r r}{u} - \delta_r. \end{cases} \tag{3.10}$$

Substitute Equations (3.10) and (3.9) into (3.8), the state-space form of linear 2-DoFs lateral dynamics model can be obtained, which was shown in Equation (3.4). For the convenience of the readers, it is copied here:

$$\dot{x}(t) = Ax(t) + B_1 w(t) + B_2 u(t), \tag{3.11}$$

where:

$$x(t) = [\beta \ r]^T, B_2 = \left[0 \ -\frac{1}{I_z}\right]^T, A = \begin{bmatrix} \dfrac{c_f + c_r}{mu} & \dfrac{l_f c_f - l_r c_r}{mu^2} - 1 \\ \dfrac{l_f c_f - l_r c_r}{I_z} & \dfrac{l_f^2 c_f + l_r^2 c_r}{I_z u} \end{bmatrix}, B = \begin{bmatrix} -\dfrac{c_f}{mu} & -\dfrac{c_r}{mu} \\ -\dfrac{l_f c_f}{I_z} & \dfrac{l_r c_r}{I_z} \end{bmatrix}, \tag{3.12}$$

where $w(t) = [\delta_f \ \delta_r]^T$ is defined as disturbance, and $u(t) = M$ is defined as control input. The vehicle dynamics model shown in Equation (3.11) does not consider the parametric uncertainties. Actually, during the motion of UGCs, some vehicle parameters may vary a lot and the nonlinearities may occur, which induces unmodeled uncertainties and disturbances. These uncertainties and disturbances may degenerate the controller's performance. Actually, for the military vehicle, the parametric uncertainties and disturbances are much more serious than that of the civilian passenger car due to the bad working environments. Therefore, if the uncertainties and the disturbances of the UGC are not considered during the controller design, the performance of the UGC will be greatly degenerated.

3.3.2   INVOLVEMENT OF THE UNMODELED UNCERTAINTIES AND DISTURBANCES

In the view of practice, the major uncertainties and disturbances of the UGC include: (1) the uncertainties of vehicle mass m and yaw inertia Iz, which occurs when the carrying load of the UGV changes. For example, if the robot and the rotorcrafts of the UGC are sent out, the vehicle mass m and yaw inertia I_z will consequently be changed; (2) the uncertainties of tire cornering stiffness c_f and c_r, which may occur when the UGC is operating in large lateral acceleration conditions and the tires are operating in nonlinear regions. For example, if the UGC is operating on the uneven road in off-road conditions, the tire vertical load will be greatly changed, and consequently the tire cornering stiffness c_f and c_r will be greatly changed; and (3) due to the application of independent steer techniques, the difference of the steering angles between four wheels may occur because of the servomotor control lag or error. This can also be assumed as the uncertainties of the tire cornering stiffness. For example, if the UGC is working in DASM or DMSM, the steer angles between four wheels may be not identical due to the control error. Therefore, the uncertainties of m, I_z, c_f, and c_r should be taken into the consideration. These uncertainties and disturbances are illustrated in Figure 3.8.

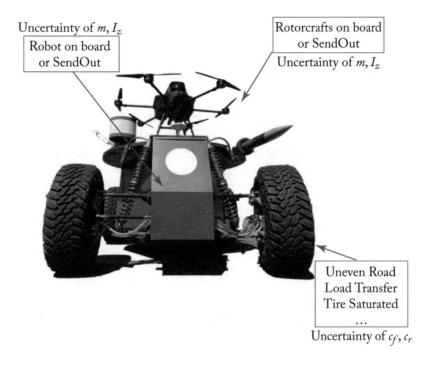

Figure 3.8: Illustration of different kinds of uncertainties.

Assume the uncertain m, I_z is bounded by their maximum value m_{max}, I_{max} and minimum value m_{min}, I_{min}. m_{max} and I_{max} is the vehicle mass and yaw inertia when the carrying load is maximum. m_{min} and I_{min} is the vehicle mass and yaw inertia when the carrying load is zero. The uncertain varying value of m and I_z can be described [22]:

$$\begin{cases} \dfrac{1}{m} = M_1 \dfrac{1}{m_{min}} + M_2 \dfrac{1}{m_{max}} \\[2mm] \dfrac{1}{I_z} = N_1 \dfrac{1}{I_{min}} + N_2 \dfrac{1}{I_{max}} \end{cases} \quad , \tag{3.13}$$

where parameters M_1, M_2, N_1, and N_2 are defined as:

$$\begin{cases} M_1 = \dfrac{1/m - 1/m_{max}}{1/m_{min} - 1/m_{max}} \\[3mm] M_2 = \dfrac{1/m_{min} - 1/m}{1/m_{min} - 1/m_{max}} \end{cases} \begin{cases} N_1 = \dfrac{1/I_z - 1/I_{max}}{1/I_{min} - 1/I_{max}} \\[3mm] M_2 = \dfrac{1/I_{min} - 1/I_z}{1/I_{min} - 1/I_{max}} \end{cases} \tag{3.14}$$

Define $h_i = M_k N_j$, where $i = 1, 2, 3, 4$, $k = 1, 2$ and $j = 1, 2$. It can be calculated that $\Sigma_i h_i = 1$.

The system in Equation (3.11) with uncertain m and I_z can be described as:

$$\dot{x}(t) = \sum_{i=1}^{4} h_i \left[A_i x(t) + B_{1i} w(t) + B_{2i} u(t) \right], \tag{3.15}$$

where A_i, B_{1i}, and B_{2i} can be obtained by replacing m and I_z with m_{min} and m_{max} and I_{min} and I_{max}, respectively.

To involve the uncertainties of the tire cornering stiffness, the real tire force should be described by nominal value cf, cr, and uncertain parts $\Delta c_f, \Delta c_r$:

$$\begin{cases} F_{yf} = (c_f + \Delta c_f) \, \alpha_f \\ F_{yr} = (c_r + \Delta c_r) \, \alpha_r. \end{cases} \tag{3.16}$$

Considering Equations (3.15) and (3.16) yields:

$$\dot{x}(t) = \sum_{i=1}^{4} h_i \left[(A_i + \Delta A_i) \, x(t) + (B_{1i} + \Delta B_{1i}) \, w(t) + B_{2i} u(t) \right]. \tag{3.17}$$

Define:

$$\begin{cases} A_0 = \sum_{i=1}^{4} h_i A_i, \; B_{10} = \sum_{i=1}^{4} h_i B_{1i}, \; B_{20} = \sum_{i=1}^{4} h_i B_{2i} \\ \Delta A = \sum_{i=1}^{4} h_i \Delta A_i, \; \Delta B_1 = \sum_{i=1}^{4} h_i \Delta B_{1i}, \end{cases} \tag{3.18}$$

where $\Delta A_i, \Delta B_{1i}$ represent the uncertainties caused by the uncertain parts Δc_f and Δc_r.

Due to the error caused by motor torque control, the uncertainty of active yaw moment should also be considered, which is defined as ΔB_2. Substitute Equation (3.18) into Equation (3.17), and then we have:

$$\dot{x}(t) = (A_0 + \Delta A)x(t) + (B_{10} + \Delta B_1)w(t) + (B_{20} + \Delta B_2)u(t). \tag{3.19}$$

In this book, we assume all the uncertainties can be bounded by the norm bounded approach. ΔA_i can be described as $\Delta A_i = DFE_i$, where D and E_i are constant known matrices, and F is unknown matrix, which is bounded by $FF^T \le I$. Therefore, we have $\Delta A = DFE_1$, where $E_1 = \sum_i h_i E_i$. Similarly, we have $\Delta B_1 = DFE_3$ and $\Delta B_2 = DFE_2$. D, E_1 and E_2 are also constant known matrices, and F is unknown matrix, which is bounded by $FF^T \le I$.

3.4   CHASSIS ROBUST YAW MOMENT CONTROL IN HIGH-SPEED CONDITIONS

3.4.1   DEFINITION OF THE TARGET CLOSED-LOOPED POLES LOCATIONS

It has been mentioned that the yaw moment will be applied in high-speed conditions to improve the handling stability of the UGC. As mentioned above, the UGC is designed inherently static unstable to improve the overall configuration flexibility with the application of CCV concept. Therefore, if the yaw moment control is not applied, the UGC will lose the stability, as Figure 3.9 shows. As discussed above, the DYC approach has been widely used in the passenger car field to improve the vehicle lateral stability with the application of the yaw moment provided by the independent motors. However, DYC is not applied for UGC, and a novel chassis yaw moment control approach is applied for UGC.

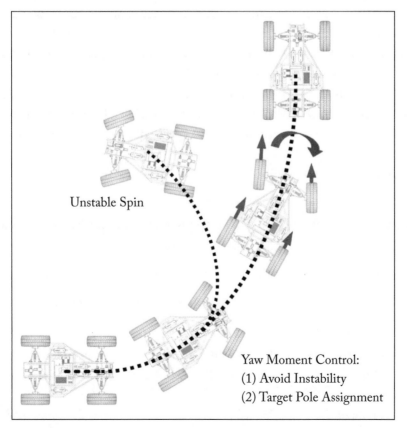

Figure 3.9: Illustration of the yaw moment control-based pole assignment technique.

In DYC's principle, the ideal vehicle handling behavior is to track the response of 2-DoFs dynamic model with inherent understeer characteristics. In other words, the ideal response of the controller is calculated by the predefined 2-DoFs linear dynamics model. It is according to the safety consideration for a passenger car. As is widely known, the response of the 2-DoFs linear dynamic model represents the understeer and stable response of the vehicle. However, it is not suitable for the application of military UGV. At first, although the safety is the most important concern for the passenger car, it is not for the military vehicle. For the military vehicle, the agility and mobility is much more important than the safety concern. Therefore, the ideal vehicle behavior of the military UGV should not be represented by the response of the 2-DoFs linear vehicle dynamic model. Second, the DYC's approach is not flexible enough to satisfy different handling performance demand of the military UGV in different missions. As discussed above, the ideal behavior is stable, which is defined by the 2-DoFs model. Therefore, the designers are not able to easily adjust the control target to adjust the vehicle's handling performance. However, a military UGV's handling performance demand may vary with different missions. For example, maybe the mobility and agility are the most important concerns in some missions, and maybe the stability is the most important concern in certain other missions. Therefore, the control approach should be flexible enough to enable the designers to adjust the target handling performance according to different missions.

To this end, a yaw moment control approach based on the pole assignment technique is proposed and used for the UGC. The control target is to use the yaw moment provided by the in-wheel motors to place the closed-looped poles of the vehicle lateral dynamics system to the target zone. In addition, the target zone can be adjusted by the designers to satisfy different handling performance requirements.

At first, the yaw moment feedback control law can be constructed as:

$$u(t) = Kx(t), \tag{3.20}$$

where K is the state-feedback gain matrix to be designed. Substituting Equation (3.20) into Equation (3.19), the closed-looped dynamics system is obtained:

$$\dot{x}(t) = [(A_0 + \Delta A) + (B_{20} + \Delta B_2)K]x(t) + (B_{10} + \Delta B_1)\,w(t). \tag{3.21}$$

Therefore, the control objective is to determine the feedback control matrix K, such that the closed-looped system in Equation (3.21) is asymptotically stable and the closed-looped poles are placed in the target zone against the uncertainties.

The target zone of the closed-looped poles of the vehicle lateral dynamics model should be described and determined. The linear matrix inequalities (LMI) region is utilized to describe the shape of the target zone. For an LMI region $\mathcal{D}$ and a system $\dot{x} = \bar{A}x$, if $\lambda(\bar{A}) \subseteq \mathcal{D}$, where λ represents the eigenvalues of the system, and then the system can be defined $\mathcal{D}$-stable. In other words, the control objective can be described as making the system in Equation (3.21) $\mathcal{D}$-stable. An LMI region is any subset $\mathcal{D}$ of the complex plane that can be defined as [80]:

$$\mathcal{D} = \{s \in C : \Gamma + s\, \Upsilon + \bar{s}\, \Upsilon^T < 0\}, \tag{3.22}$$

where C denotes the complex set; Γ is a real symmetric matrix and assume $\Gamma \in R^{m \times m}$; Υ is a real symmetric matrix and assume $\Upsilon \in R^{m \times m}$. The characteristics function of LMI region $\mathcal{D}$ is defined as:

$$f_{\mathcal{D}}(s) = \Gamma + s\, \Upsilon + \bar{s}\, \Upsilon^T. \tag{3.23}$$

The disk centered at $(-q, 0)$ with radius as r is typical LMI region in control application, which will be used to describe the zone of target closed-looped poles. It can be described as:

$$\mathcal{D}_{q,r} = \{s \in C \mid (s + q)(\bar{s} + q) - r^2 < 0\}. \tag{3.24}$$

Its characteristics function can be described as:

$$f_{\mathcal{D}_{q,r}}(s) = \begin{bmatrix} -r & q+s \\ q+\bar{s} & -r \end{bmatrix}. \tag{3.25}$$

An illustration of the disk zone of target closed-looped poles is shown in Figure 3.10. The selection of the target poles locations is very important, which determines the target handling performance of the vehicle. The most important is to select the damping ratio ξ and undamped natural frequency ω_n. If a certain disk zone is selected, the value of ξ and ω_n can be limited in a range. As Figure 3.10 shows, if $\lambda \subseteq \mathcal{D}_{q,r}$, then we have $\xi \geq \sqrt{1 - (r/q)^2}$ and $q - r \leq \xi\omega_n \leq q + r$. Therefore, the disk zone can be determined according to the requirement of ξ and ω_n. For example, if the minimum value of ξ is given as a and the minimum value of $\xi\omega_n$ is given as b, then the target disk zone can be described by:

$$\begin{cases} \theta = \arccos a \\ q = b/(1 - \sin\theta) \\ r = b\sin\theta / (1 - \sin\theta) . \end{cases} \tag{3.26}$$

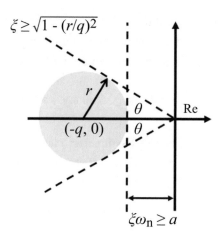

Figure 3.10: Illustration of the target closed-looped poles locations.

The designer can select the target poles' locations according to the particular demand of different missions, which is a unique advantage of this approach. Figure 3.11 shows several examples of target disk zones in six different cases ($a = 0.8$, $b = 4$; $a = 0.8$, $b = 5$; $a = 0.8$, $b = 6$; $a = 0.7$, $b = 4$; $a = 0.7$, $b = 5$; $a = 0.7$, $b = 6$). It can be concluded that when b increases (a is constant), the target zone circle moves to left and the radius becomes larger. When a increases (b constant), the target zone circle moves to right and the radius becomes smaller. Since the value of b represents the natural frequency and the static margin, the value of a represents the damping ratio. The designer can select the desired poles locations according to the particular handling performance demand. Generally, following points should be noticed during the selection of the target poles locations.

1. If this approach is applied for the mission with the requirement of better agility and quicker transient response, then the target poles' locations should be selected with higher frequency, smaller static margin, and appropriate damping ratio.

2. If this approach is applied for the mission with the requirement of better stability, then the target poles' locations should be selected with higher static margin.

3. The target poles' locations can be designed as a function of the vehicle speed to adjust the handling performance in different speed condition. For example, higher agility may be wanted in low-speed conditions, and higher stability may be wanted in high-speed conditions.

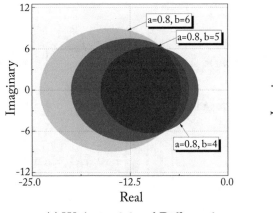

(a) With a as 0.8 and Different b

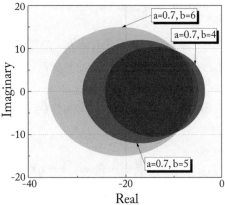

(a) With a as 0.7 and Different b

Figure 3.11: Target poles location examples.

3.4.2   DESIGN OF THE ROBUST POLE ASSIGNMENT CONTROLLER

Given the description of the target poles zone, the controller can be designed accordingly. For a dynamics system and a given zone $\mathcal{D}$, the system is $\mathcal{D}$-stable if and only if there is asymmetric positive matrix Y such that:

$$M_{\mathcal{D}}(\bar{A}, Y)\, \Gamma \otimes Y + \Upsilon \otimes \bar{A}Y + \Upsilon^T \otimes \bar{A}^T Y < 0. \tag{3.27}$$

Assume a symmetric matrix H, and Θ, Ψ are real matrices with proper dimensions. Assume $\Lambda^T \Lambda < \mathrm{I}$, and then the following condition:

$$H + \Theta \Lambda \Psi + \Psi^T \Lambda^T \Theta^T < 0 \tag{3.28}$$

holds if and only if there exists a positive scalar $\varepsilon > 0$ such that:

$$H + \varepsilon \Theta \Theta^T + \varepsilon^{-1} \Psi^T \Psi < 0. \tag{3.29}$$

Therefore, for a disk LMI region (q, r), the dynamics system in Equation (3.21) is $\mathcal{D}$-stable only if there is a positive symmetric matrix Y such that:

$$\begin{bmatrix} -r\,Y & (A_0 + \Delta A + B_{20}\,K + \Delta B_2 K)Y + qY \\ (A_0 + \Delta A + B_{20}\,K + \Delta B_2 K)Y + qY & -rY \end{bmatrix} < 0. \tag{3.30}$$

Substituting ΔA, ΔB_1, and ΔB_2 into Equation (3.30) yields:

$$\begin{bmatrix} -rY & (A_0 + B_{20}K)Y + qY \\ (A_0 + B_{20}K)Y + qY & -rY \end{bmatrix} + \begin{bmatrix} 0 & \mathcal{D}F(E_1 + E_3K)Y \\ \mathcal{D}F(E_1 + E_3K)Y & 0 \end{bmatrix} < 0. \quad (3.31)$$

After manipulation, Equation (3.31) can be changed to:

$$\begin{bmatrix} -rY & (A_0 + B_{20}K)Y + qY \\ (A_0 + B_{20}K)Y + qY & -rY \end{bmatrix} + \begin{bmatrix} \mathcal{D} \\ 0 \end{bmatrix} F[0 \quad E_1Y + E_2KY]$$

$$+ [0 \quad E_1Y + E_2KY]^T F^T \begin{bmatrix} \mathcal{D} \\ 0 \end{bmatrix}^T < 0. \quad (3.32)$$

Define parameter Q to simplify the expression:

$$Q = \begin{bmatrix} -rY & (A_0 + B_{20}K)Y + qY \\ (A_0 + B_{20}K)Y + qY & -rY \end{bmatrix}. \quad (3.33)$$

According to above lemma shown in Equations (3.28) and (3.29), there is only one positive scalar ε_0 exists satisfying the following relationship:

$$\varepsilon_0^2 [0 \quad E_1Y + E_3KY]^T [0 \quad E_1Y + E_3KY] + \varepsilon_0 Q + \begin{bmatrix} \mathcal{D} \\ 0 \end{bmatrix} \begin{bmatrix} \mathcal{D} \\ 0 \end{bmatrix}^T < 0. \quad (3.34)$$

After manipulation, Equation (3.34) can be changed to:

$$\begin{bmatrix} \mathcal{D}\mathcal{D}^T & 0 \\ 0 & \varepsilon_0^2 [(E_1Y + E_3K)Y]^T [(E_1 + E_3K)Y] \end{bmatrix} + \varepsilon_0 Q < 0. \quad (3.35)$$

To solve the LMI, the Schur complement theorem should be introduced. Assume matrix $S \in R^{m \times m}$, which can be described as:

$$S = \begin{bmatrix} S_{11} & S_{12} \\ S_{21} & S_{22} \end{bmatrix}. \quad (3.36)$$

The following three inequalities are identical:

$$\begin{cases} S < 0 \\ S_{11} < 0, S_{22} - S_{12}^T S_{11}^{-1} S_{12} < 0 \\ S_{22} < 0, S_{11} - S_{12}^T S_{22}^{-1} S_{12} < 0. \end{cases} \quad (3.37)$$

Therefore, Equation (3.35) can be changed to:

$$\begin{bmatrix} \varepsilon_0 rY & \mathcal{D} & \varepsilon_0[qI + (A_0 + B_{20}K)]Y & \varepsilon_0[(E_1 + E_3K)Y]^T \\ \mathcal{D} & -I & 0 & 0 \\ \varepsilon_0[qI + (A_0 + B_{20}K)]Y & 0 & -\varepsilon_0 rY & 0 \\ \varepsilon_0[(E_1 + E_3K)Y]^T & 0 & 0 & -I \end{bmatrix} < 0. \quad (3.38)$$

Define $Z = \varepsilon_0 Y$, $P = KZ$, and Equation (3.39) can be rewritten as:

$$\begin{bmatrix} -rZ & \mathcal{D} & qZ + (A_0Z + B_{20}P) & (E_1Z + E_3P)^T \\ \mathcal{D} & -I & 0 & 0 \\ qZ + (A_0Z + B_{20}P) & 0 & -rZ & 0 \\ (E_1Z + E_3P)P]^T & 0 & 0 & -I \end{bmatrix} < 0. \quad (3.39)$$

By using Matlab LMI solving toolbox, the feedback controller gain can be obtained as $K = PZ^{-1}$. The desired yaw moment value can be calculated with the feedback of yaw velocity and side slip angle.

3.5   CHASSIS ROBUST YAW MOMENT CONTROL IN DMSM

3.5.1   CONTROL OBJECTIVE DEFINITION IN DMSM

With the AWIS technique, DMSM can be achieved to control the UGC to move laterally or diagonally from one point directly to another point without yaw motion, which can greatly improve the maneuverability and path-tracking performance of the UGC [23]. In most of the UGV's missions, such as search, rescue, cargo delivery, and other maneuvers in narrow spaces, the DMSM can significantly improve the performance and reduce the length of time.

An illustration example in lane change maneuver is shown in Figure 3.12 to show how the yaw moment controller is applied in DMSM. As Figure 3.12 illustrates, DMSM allows the UGC to finish the lane change without steering motion and keep heading to the same direction. During the lane change phase, the steering angles of all the wheels are controlled to be the same direction. However, the path deviation or other instability phenomenon may occur because of the disturbances, such as the cross wind, the difference of the friction coefficient between each wheels or the difference of the steering angles between each wheels. In these cases, the sideslip angle of the vehicle may change and the yaw motion may occur. Therefore, the active yaw moment provided by the independent motors can be applied to track the desired sideslip angle and minimize the yaw velocity, which aims at assuring the desired behavior in DMSM against the disturbances.

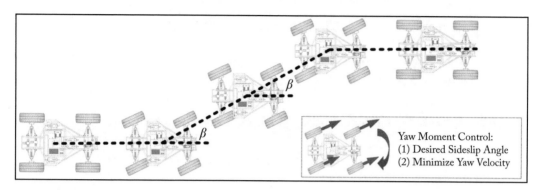

Figure 3.12: Illustration of yaw moment control in lane change maneuver in DMSM.

Therefore, the desired value of sideslip angle and yaw velocity in DMSM can be described as:

$$\begin{cases} \beta_d(t) = \delta \\ r_d(t) = 0. \end{cases} \tag{3.40}$$

The steering angle of the front and rear wheels in DMSM is assumed to be δ. To track the desired path without yaw motion, the desired sideslip angle should be equal to the steering angle.

The yaw moment feedback control law can also be constructed as this form:

$$u(t) = Kx(t), \tag{3.41}$$

where K is the state-feedback gain matrix to be designed. Therefore, the closed-looped dynamics system is:

$$\dot{x}(t) = [(A_0 + \Delta A) + (B_{20} + \Delta B_2)K]x(t) + (B_{10} + \Delta B_1)w(t). \tag{3.42}$$

According to Equation (3.40), define yaw velocity and sideslip angle as control outputs:

$$\begin{cases} z_1(t) = \beta_1(t) = C_1x(t) \\ z_2(t) = r(t) = C_2x(t), \end{cases} \tag{3.43}$$

where $C_1 = [1\ 0]^T$ and $C_1 = [0\ 1]^T$.

By using H infinity approach, the control goal for $C_1(t)$ can be defined as:

$$\|z_1(t)\|_\infty < \gamma_1 \|w(t)\|_2, \tag{3.44}$$

where γ_1 is a given small number, and:

$$\|z_1(t)\|_\infty = \sup_{t \in [0, \infty]} \sqrt{z_1^T(t)z_1(t)}. \tag{3.45}$$

The H infinity gain is selected as the performance measurement for $C_2(t)$. The control goal can be defined as:

$$\|T_{z2w}\|_\infty = \sup_{\|w(t)\|_2 \neq 0} \frac{\|z_2(t)\|_2}{\|w(t)\|_2} < \gamma_2, \qquad (3.46)$$

where γ_2 is a given small number, and:

$$\begin{cases} \|z_2(t)\|_2^2 = \int_0^\infty z_2(t)\, z_2(t) dt \\ \|w(t)\|_2^2 = \int_0^\infty w^T(t)\, w(t) dt . \end{cases} \qquad (3.47)$$

Therefore, the control goal of the yaw moment controller in DMSM can be described as to make the closed-looped system in Equation (3.42) asymptotically stable, and such that the performance shown in (3.44) and (3.46) is achieved.

3.5.2   DESIGN OF THE ROBUST H INFINITY YAW MOMENT CONTROLLER IN DMSM

In order to obtain the performance shown in Equation (3.46), define a Lyapunov function for the system in Equation (3.42) as:

$$V(x(t)) = x^T(t)\, P_1 x(t), \qquad (3.48)$$

where P_1 is a positive matrix. Differentiating Equation (3.48) yields:

$$\dot{V}(x(t)) = \dot{x}^T(t)\, P_1 x(t) + x^T(t)\, P_1 \dot{x}(t). \qquad (3.49)$$

Substitute Equation (3.42) into Equation (3.49):

$$\dot{V}(x(t)) = [(A_0 + B_{20}K + \Delta A + \Delta B_2 K)x(t) + (B_{10} + \Delta B_1)\, w(t)]^T P_1 x(t) + \\ x^T(t)\, P_1[(A_0 + B_{20}K + \Delta A + \Delta B_2 K)x(t) + (B_{10} + \Delta B_1)\, w(t)]. \qquad (3.50)$$

Substitute ΔA, ΔB_1, and ΔB_2 into Equation (3.50), and then we have:

$$\dot{V}(x(t)) = x^T(t)\, [(A_0 + B_{20}K)^T P_1 + P_1(A_0 + B_{20}K) + P_1 DF(E_1 + E_2) + (E_1 + E_2)^T F^T D^T P_1|]\, x(t) \\ x^T(t)\, P_1(B_{10} + DFE_3)\, w(t) + w^T(t)\, (B_{10} + DFE_3)^T P_1 x(t). \qquad (3.51)$$

According to the lemma shown in Equations (3.28) and (3.29), the following can be known from Equation (3.51):

$$\dot{V}(x(t)) \leq x^T (t) \left[(A_0 + B_{20}K)^T P_1 + P_1 (A_0 + B_{20}K) + \varepsilon_1 P_1 DD^T P_1 + \varepsilon_2 P_1 DD^T P_1 + \varepsilon_1^{-1} E_1^T E_1 + \varepsilon_2^{-1} E_2^T \right.$$
$$\left. E_2 \right] x(t) + x^T P_1 B_{10} w(t) + w^T B_{10} P_1 x(t) + \varepsilon_3 x^T (t) P_1 DD^T P_1 x(t) + \varepsilon_3^{-1} w^T (t) E_3^T E_3 w(t). \quad (3.52)$$

Add $z_2^T(t) z_2 (t) - \gamma_2^2 w^T (t) w (t)$ to both sides of Equation (3.52), and then we have:

$$\dot{V}(x(t)) + z_2^T(t) z_2 (t) - \gamma_2^2 w^T (t) w (t) \leq \begin{bmatrix} x(t) \\ w(t) \end{bmatrix}^T \Lambda \begin{bmatrix} x(t) \\ w(t) \end{bmatrix}, \quad (3.53)$$

where:

$$\Lambda = \begin{bmatrix} (A_0 + B_{20}K)^T P_1 + P_1 (A_0 + B_{20}K) & & \\ + (\varepsilon_1 + \varepsilon_2 + \varepsilon_3) P_1 DD^T P_1 + (\varepsilon_1^{-1} + \varepsilon_2^{-1}) E_2^T E_2 & C_2^T C_2 & P_1 B_{10} \\ C_2^T C_2 & -I & 0 \\ P_1 B_{10} & 0 & -\gamma_2^2 I + \varepsilon_3 E_3^T E \end{bmatrix}. \quad (3.54)$$

It can be known that if $\Lambda < 0$, the left terms of Equation (3.53) will be lower than zero, and then $V(x(t)) < 0$. Therefore, the system will be quadratically stable. Premultiply and postmultiply Λ by diag $(P^{-1}I)$ and its transpose, respectively. The condition $\Lambda < 0$ is equivalent to:

$$\Lambda = \begin{bmatrix} P_1^{-1} (A_0 + B_{20}K)^T + (A_0 + B_{20}K) P_1^{-1} + (\varepsilon_1 + \varepsilon_2 + \varepsilon_2) DD^T & & \\ + \varepsilon_1^{-1} P_1^{-1} E_1^T E_1 P_1^{-1} + \varepsilon_2^{-1} P_1^{-1} E_2^T E_2 P_1^{-1} & P_1^{-1} C_2^T & B_1 \\ P_1^{-1} C_2^T & -I & 0 \\ B_1 & 0 & -\gamma_2^2 I + \varepsilon_3^{-1} E_3^T E_3 \end{bmatrix}. \quad (3.55)$$

Defining $Z_1 = P_1^{-1}$, $Q_1 = K Z_1$, Equation (3.55) can be changed to:

$$\Lambda = \begin{bmatrix} (A_0 Z_1 + B_{20} Q_1)^T + (A_0 Z_1 + B_{20} Q_1) + (\varepsilon_1 + \varepsilon_2 + \varepsilon_3) DD^T & & & \\ + \varepsilon_1^{-1} Z_1 E_1^T E_1 Z_1 + \varepsilon_2^{-1} Z_1 E_2^T E_2 Z_1 & Z_1 C_2^T & B_{10} & 0 \\ Z_1 C_2^T & -I & 0 & 0 \\ B_{10} & 0 & -\gamma_2^2 & E_3^T \\ 0 & 0 & E_3^T & -\varepsilon_1 \end{bmatrix} < 0 \quad (3.56)$$

According to Schur complement, the following can be obtained:

$$\begin{bmatrix} (A_0Z_1 + B_{20}Q_1)^T + (A_0Z_1 + B_{20}Q_1) \\ + (\varepsilon_1 + \varepsilon_2 + \varepsilon_3)DD^T & Z_1E_1^T & Z_1E_2^T & Z_1C_2^T & B_{10} & 0 \\ Z_1E_1^T & -\varepsilon_1I & 0 & 0 & 0 & 0 \\ Z_1E_2^T & 0 & -\varepsilon_2I & 0 & 0 & 0 \\ Z_1C_2^T & 0 & 0 & -I & 0 & 0 \\ B_{10} & 0 & 0 & 0 & -\gamma_2^2 & E_3^T \\ 0 & 0 & 0 & 0 & E_3^T & -\varepsilon_3I \end{bmatrix}. \tag{3.57}$$

To obtain the performance in Equation (3.44), the following lemma is introduced. Consider the system in state-space from:

$$\begin{cases} \dot{X} = AX + BU \\ Z = CX \end{cases}. \tag{3.58}$$

Given a positive scalar γ, and then following condition

$$\|Z\|_\infty < \gamma \, \|W\|_2 \tag{3.59}$$

holds if and only if there exists a positive matrix P2 such that:

$$\begin{cases} AP_2 + P_2A^T + BB^T < 0 \\ CP_2C^T < \gamma^2 I \end{cases}. \tag{3.60}$$

The following required condition can be obtained according to the first part of Equation (3.60) and Equation (3.42):

$$[(A_0 + \Delta A) + (B_{20} + \Delta B_2)K]P_2 + P_2[(A_0 + \Delta A) + (B_{20} + \Delta B_2)K]^T + (B_{10} + \Delta B_1)(B_{10} + \Delta B_1)^T < 0. \tag{3.61}$$

Substitute ΔA, ΔB_1, and ΔB_2 into Equation (3.61), and then we have:

$$(A_0 + B_{20}K)P_2 + P_2(A_0 + B_{20}K)^T + [D \ 0] F \begin{bmatrix} (E_1 + E_2K)P_2 + E_3 B_{10}^T \\ 0 \end{bmatrix} \\ + \begin{bmatrix} (E_1 + E_2K)P_2 + E_3 B_{10}^T \\ 0 \end{bmatrix}^T F^T [D \ 0]^T < 0. \tag{3.62}$$

According to the lemma shown in Equations (3.28) and (3.29), Equation (3.62) can be changed to:

$$\varepsilon_4[(A_0 + B_{20} K) P_2 + P_2(A_0 + B_{20} K)^T + B_{10}B_{10}^T + [D\ 0]\ [D\ 0]^T$$

$$+ \varepsilon_4^2 \begin{bmatrix} (E_1 + E_2K)P_2 + E_3\ B_1^T \\ 0 \end{bmatrix}^T \begin{bmatrix} (E_1 + E_2K)P_2 + E_3\ B_1^T \\ 0 \end{bmatrix}. \tag{3.63}$$

According to the Schur complement, it can be changed to:

$$\begin{bmatrix} \varepsilon_4[(A_0 + B_{20}K) P_2 + P_2(A_0 + B_{20}K)^T + B_{10}B_{10}^T + DD^T & \varepsilon_4[(E_1 + E_2K) P_2 + E_3B_{10}^T]^T \\ \varepsilon_4[(E_1 + E_2K) P_2 + E_3B_{10}^T] & -I \end{bmatrix} < 0. \tag{3.64}$$

Defining $Z_2 = \varepsilon_4P_2^{-1}$ and $Q_2 = KZ_2$, it can be changed to:

$$\begin{bmatrix} [(A_0Z_2 + B_{20}Q_2) + (A_0Z_2 + B_{20}Q_2)^T + \varepsilon_4\ B_{10}B_{10}^T] + DD^T & [(E_1Z_2 + E_2Q_2)\ P_2 + \varepsilon_4\ E_3B_{10}^T]^T \\ (E_1Z_2 + E_2Q_2) + \alpha E_3B_{10}^T & -I \end{bmatrix} < 0. \tag{3.65}$$

To obtain the performance in Equation (3.44), the following required condition can be obtained according to the second par of Equation (3.60) and Equation (3.42):

$$\begin{bmatrix} -P_2^{-1} & C_1^T \\ C_1 & -\gamma_1^2 I \end{bmatrix} > 0. \tag{3.66}$$

Premultiply and postmultiply it by diag (P2 I) yields:

$$\begin{bmatrix} -P_2 & C_1^TP_2 \\ P_2C_1 & -\gamma_1^2 I \end{bmatrix} > 0. \tag{3.67}$$

Define $Z_3 = P_2$, and the Equation (3.67) can be changed to:

$$\begin{bmatrix} -Z_3 & C_1^TZ_3 \\ Z_3C_1 & -\gamma_1^2 I \end{bmatrix} > 0. \tag{3.68}$$

The state feedback gain K can be calculated by Matlab LMI solving toolbox according to the LMI (3.57), (3.65), and (3.68).

3.6   ROBUST CHASSIS CONTROL WITH CONSIDERATION OF TIME DELAY

3.6.1   DESCRIPTION OF THE TIME DELAY OF THE CONTROL SYSTEM

In the above sections, the robust chassis yaw moment controllers in ASM and DMSM are designed to improve the handling stability and the path tracking performance, respectively. However, there is another unique problem for the full x-by-wire and remote control UGV, which is the delay. The delay problem is very common in the control engineering practice. For the x-by-wire vehicle, the actuators, such as driven motors and steer servomotors, receive the control signal from the controller through CAN bus. The delay may occur during the sending and receiving of each data pack, which can be called "controller to actuator" delay. In addition, for the UGC, the delay phenomenon may be much worse than above. Figure 3.13 shows the illustration of all the kinds of delay of the UGC. It can be seen that there are other kinds of delays besides the "controller to actuator" delay. According to the remote control strategy of the UGC, the human operator needs to observe the image feedback by the cameras carried on the UGC to determine the next actions. There is a large delay of the image feedback transmission. After the human operator receives the image through the monitor, he needs a little time to think and react, which causes the human delay. After the human operator finishes the manipulation of the command station, there is delay of the tele-transmission system (from the command station to the ECU controller). After the controller receives the control signal, the control signal will be transmitted to the actuator. However, there is still a delay between the controller and the actuator.

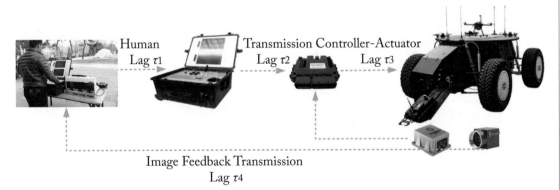

Figure 3.13: Illustration of delays when the UGC is working.

According to Figure 3.13, the total delay of the network can be described as:

$$\tau(t) = \tau_1(t) + \tau_2(t) + \tau_3(t) + \tau_4(t). \tag{3.69}$$

Assume the time delay $\tau(t)$ is bounded in range $(0, t_d)$. The state variable x(t) is collected by the sensor, and then it will be transmitted to the ECU control to conduct the dynamics control. Therefore, defining the state variables received by the sensor at sampling time k is:

$$\hat{x}(t_k) = x(t_k - \tau(t)). \tag{3.70}$$

The yaw moment feedback control law can be reconstructed as:

$$u = K\hat{x}(t_k) = Kx(t - \tau(t)). \tag{3.71}$$

The closed-looped dynamics system with the yaw moment feedback can be obtained:

$$x(t) = (A_0 + \Delta A)x(t) + (B_{20} + \Delta B_2)Kx(t - \tau(t)) + (B_{10} + \Delta B_1)w(t). \tag{3.72}$$

3.6.2 DESIGN OF THE ROBUST CONTROLLER CONSIDERING THE TIME DELAY

In this section, the DMSM will be taken as example to handle the time delay problem. Therefore, the control objective is the same as Equations (3.43)–(3.46) shown. The following lemma is given to handle the time delay problem: Assume the positive matrices Σ, Θ and $\Psi \in \mathbf{R}^{n \times n}$, and the following relationship always holds:

$$-2\Theta^T \Psi \le \Theta^T \Sigma^{-1}\Theta + \Psi^T \Sigma \Psi. \tag{3.73}$$

Consider system in the state-space form as:

$$\begin{cases} \dot{X} = AX + B_2KX(t - t_d(t)) + B_1W \\ Z = CX. \end{cases} \tag{3.74}$$

The system shown in Equation (3.74) is stable if and only if there exist positive matrix P and Q satisfying the following relationship:

$$\begin{cases} \begin{bmatrix} A^TP + PA + Q & PB_2K \\ * & -(1-t_d)Q \end{bmatrix} < 0, \\ CPC^T < \gamma^2 IA \end{cases} \tag{3.75}$$

where γ is a given positive scalar.

According to the definition of the derivative, there is:

$$x(t) - x(t - \tau) = \int_{t-\tau}^{t} \dot{x}(s)ds. \tag{3.76}$$

The closed-looped system in Equation (3.72) can be rewritten as:

$$\dot{x}(t) = [A_0 + \Delta A + (B_{20} + \Delta B_2)K] - (B_{20} + \Delta B_2)K \int_{t-\tau}^{t} \dot{x}(s)ds + (B_{10} + \Delta B_1)w(t). \tag{3.77}$$

In order to discuss the stability of the control system, define the system's Lyapunov function as:

$$\begin{cases} V(x(t)) = xV_1(x(t)) + V_2(x(t)) + V_3(x(t)) \\ V_1(x(t)) = x(t)^T Px(t) \\ V_2(x(t)) = \int_{t-t_d}^{t} x^T(s) Rx(s)ds \\ V_3(x(t)) = \int_{-t_d}^{0} \int_{t+\beta}^{t} x^T(\alpha) Qx(\alpha)d\alpha d\beta , \end{cases} \tag{3.78}$$

where P, R, and Q are symmetric positive matrices.

Differentiate the Lyapunov function, and consider the above lemmas:

$$\dot{V}(x(t)) = \dot{V}_1(x(t)) + \dot{V}_2(x(t)) + \dot{V}_3(x(t))$$
$$\leq x^T(t)\{sys[P((A_0 + \Delta A) + (B_0 + \Delta B)K] + R + t_d P(B_0 + \Delta B)KQ^{-1}K^T (B_0 + \Delta B)^TP\} x(t)$$
$$x^T(t - t_d)Rx(t - t_d) + x^T(t) Pw(t) + w^T(t)Px^T(t) + t_d\dot{x}. \tag{3.79}$$

Define:

$$\Omega^T = [x^T(t), x^T(t - t_d), x^T(t - \tau), w^T(t)]. \tag{3.80}$$

Equation (3.79) can be rewritten as:

$$\dot{V}(x(t)) = (x(t)) \leq \Omega^T(t)\Xi\Omega(t) + t_d\dot{x}^T(t)Q\dot{x}(t), \tag{3.81}$$

where:

$$\Xi = \begin{bmatrix} sys[P((A_0 + \Delta A) + (B_{20} + \Delta B_{20})K)] + R \\ \quad + t_d P_1 (B_{20} + \Delta B_{20})KQ^{-1}K^T (B_{20} + \Delta B_2)^T P & 0 & 0 & P \\ * & -R & 0 & 0 \\ * & * & 0 & 0 \\ * & * & * & 0 \end{bmatrix}. \tag{3.82}$$

Add $z_2^T(t)z_2(t) - \gamma^2 w^T(t)w(t)$ to two sides of Equation (3.81). According to the Schur complement, it can be obtained:

$$\dot{V}(x(t)) + z_2^T(t)\, z_2(t) - \gamma^2 w^T(t)w(t) \leq \Sigma^T(t)\Delta\Sigma(t), \tag{3.83}$$

where:

$$\Xi = \begin{bmatrix} sys[P((A_0 + \Delta A) + (B_{20} + \Delta B_{20})K)^T] + R & P(B_{20} + \Delta B_{20})K & 0 & P & (A_0 + \Delta A)^T \\ * & -t_d^{-1}Q & 0 & 0 & (B_{20} + \Delta B_{20})K \\ * & * & -R & 0 & 0 \\ * & * & * & -\gamma^2 I & 0 \\ * & * & * & * & -t_d^{-1}Q^{-1} \\ * & * & * & * & * \end{bmatrix}. \tag{3.84}$$

It can be known that if $\Delta < 0$, the derivative of the Lyapunov function can be negative. Substituting the description of the uncertainty into Equation (3.84), it can be obtained that:

$$\Delta = \begin{bmatrix} sys[P(A + B_{20}K)^T x(t)] + R & PB_{20}K & 0 & P & A_0^T & C^T \\ * & -t_d^{-1}Q & 0 & 0 & (B_{20}K)^T & 0 \\ * & * & -R & 0 & 0 & 0 \\ * & * & * & -\gamma^2 I & 0 & 0 \\ * & * & * & * & -t_d^{-1}Q^{-1} & 0 \\ * & * & * & * & * & -I \end{bmatrix} +$$

$$sys \begin{pmatrix} \begin{bmatrix} PD & 0 & & (E_1 + E_2 K)^T & E_1^T \\ 0 & 0 & & (E_2 K)^T & (E_2 K)^T \\ 0 & 0 & F1 & 0 & 0 \\ 0 & 0 & F2 & 0 & 0 \\ 0 & 0 & & 0 & 0 \\ 0 & 0 & & 0 & 0 \end{bmatrix} \end{pmatrix}. \tag{3.85}$$

According to the Schur complement and the above lemma, it can be known that $\Delta < 0$ is equal to $\Delta 1 < 0$, where:

$$\Delta_1 = \begin{bmatrix} \Delta_{11} & \Delta_{12} \\ * & \Delta_{22} \end{bmatrix}, \tag{3.86}$$

$$\Delta_{11} = \begin{bmatrix} sys[P(A_0 + B_{20}K)^T x(t)] + R & PB_{20}K & 0 & P & A_0^T & C^T \\ * & -t_d^{-1}Q & 0 & 0 & (B_{20}K)^T & 0 \\ * & * & -R & 0 & 0 & 0 \\ * & * & * & -\gamma^2 I & 0 & 0 \\ * & * & * & * & -t_d^{-1}Q^{-1} & 0 \\ * & * & * & * & * & -I \end{bmatrix}, \tag{3.87}$$

$$\Delta_{12} = \begin{bmatrix} \varepsilon_1 PD & 0 & (E_1 + E_2 K)^T & E_1^T \\ 0 & 0 & (E_2 K)^T & (E_2 K)^T \\ 0 & 0 & 0 & 0 \\ 0 & 0 & 0 & 0 \\ 0 & \varepsilon_2 D & 0 & 0 \\ 0 & 0 & 0 & 0 \end{bmatrix}, \tag{3.88}$$

$$\Delta_{22} = \begin{bmatrix} \xi_1 I & & & \\ & \xi_2 I & & \\ & & \xi_1 I & \\ & & & \xi_2 I \end{bmatrix}. \tag{3.89}$$

By premultiplying and postmultiplying Δ_1 by diag $(P^{-1}, P^{-1}, I, I, I, I, I, I, I, I)$, we can get the congruent transformation of the matrix. Due to the space limit, some procedures are neglected here. Define $X = P^{-1}$, $Z = KP^{-1}$, $Q_0 = P^{-1}RP^{-1}$, $Q_1 = P^{-1}QP^{-1}$, $Q_2 = Q^{-1}$, the final LMI can be obtained.

$$\begin{bmatrix} \Phi_1 & \Phi_2 \\ * & \Phi_3 \end{bmatrix} < 0, \tag{3.90}$$

where:

$$\Phi_1 = \begin{bmatrix} sys(XA_0^T + B_{20}Z) + Q_0 & B_{20}Z & 0 & I & XA_0^T & C_2^T \\ * & -t_d^{-1}Q_1 & 0 & 0 & Z^TB_{20}^T & 0 \\ * & * & -R & 0 & 0 & 0 \\ * & * & * & -\gamma^2 I & 0 & 0 \\ * & * & * & * & -t_d^{-1}Q_2 & 0 \\ * & * & * & * & * & -I \end{bmatrix}, \tag{3.91}$$

$$\Phi_2 = \begin{bmatrix} \varepsilon_1 D & 0 & (E_1X + E_2Z)^T & XE_1^T \\ 0 & 0 & (E_2Z)^T & (E_2Z)^T \\ 0 & 0 & 0 & 0 \\ 0 & 0 & 0 & 0 \\ 0 & \varepsilon_2 D & 0 & 0 \\ 0 & 0 & 0 & 0 \end{bmatrix}, \tag{3.92}$$

$$\Phi_3 = \begin{bmatrix} -\varepsilon_1 I & 0 & 0 & 0 \\ * & -\varepsilon_2 I & 0 & 0 \\ * & * & -\varepsilon_1 I & 0 \\ * & * & * & -\varepsilon_2 I \end{bmatrix}. \tag{3.93}$$

Rewriting the first LMI in Equation (3.75):

$$\begin{bmatrix} (A_0 + \Delta A)^T P + P(A_0 + \Delta A) + Q & P(B_{20} + \Delta B_{20})K \\ * & -(1 - t_d)Q \end{bmatrix} < 0. \tag{3.94}$$

Considering the description of the uncertainty, Equation (3.94) can be changed:

$$\begin{bmatrix} A_0^T P + PA_0 + R & PB_0K \\ * & -(1 - t_d)R \end{bmatrix} + sys \begin{bmatrix} PDFE_1 & P(DFE_2)K \\ * & 0 \end{bmatrix} < 0. \tag{3.95}$$

According to the above lemmas and the Schur complement, Equation (3.95) can be changed to:

$$\begin{bmatrix} A_0^T P + PA_0 + R & PB_0K & \varepsilon_3 PD & E_1^T \\ * & -(1-t_d)R & 0 & K^T E_2^T \\ * & * & -\varepsilon_3 I & 0 \\ * & * & * & -\varepsilon_3 I \end{bmatrix} < 0. \qquad (3.96)$$

By premultiplying and postmultiplying (3.96) by diag $(P^{-1}, P^{-1}, I, I, I)$, we can get the congruent transformation of the matrix, and then by defining $X = P^{-1}$, $Z = KP^{-1}$, $Q_0 = P^{-1}RP^{-1}$, the final LMI can be obtained.

$$\begin{bmatrix} A_0 X + XA_0^T + Q_0 & B_0Z & \varepsilon_3 D & XE_1^T \\ * & -(1-t_d)Y & & ZE_2^T \\ * & * & -\varepsilon_3 I & \\ * & * & * & -\varepsilon_3 I \end{bmatrix} < 0 . \qquad (3.97)$$

The control feedback gain K of the robust controller can be obtained by the LMIs (3.90) and (3.97).

3.7   TIRE FORCES DISTRIBUTION AND TIRE SLIP RATIO CONTROL

3.7.1   TIRE FORCES DISTRIBUTION BASED ON TIRE VERTICAL LOAD

As introduced above, in upper controller of the overall control block, the yaw moment will be applied to improve the handling performance, maneuverability, and mobility of the UGC. In the previous sections, only the yaw moment controllers in ASM and DMSM are detailed. The reason should be discussed here. In Sections 3.3–3.6, the yaw moment controllers in ASM and DMSM are designed, which are mainly based on robust H infinity approaches. DASM is mainly used in low-speed turning conditions to negotiate narrow spaces. In DASM, the yaw moment is supposed to help the UGC to reduce the steer radius. The yaw moment value is designed in proportion to the value of the steer angle. In other words, the larger the steer angle of the wheels, the larger yaw moment will be applied to reduce the steer radius. Therefore, the yaw moment control strategy in DASM is simple, which is neglected in this book.

After the yaw moment value is determined in the upper control, the lower controller should calculate how much traction force should be generated at each wheel. The traction force on each wheel is determined by both yaw moment and target total traction force. In this section, the longitudinal force distribution strategy will be detailed, which is very important to improve the tire adhesion margin and further enhances the handling stability. As mentioned above, the UGC is mainly used in military applications, such as in field environments. When the UGC is operating on the uneven road in field environments, the tire vertical load of each tire may vary a lot due to the road roughness. At some time, some certain tires may even lost the contact with the ground. Therefore, the appropriate tire force distribution is very important for the UGC to negotiate even road.

The basic principle of the tire longitudinal force aims at improving the overall tire adhesion capability of the tires. According to the tire friction circle concept, the tire adhesion capability is in proportion to the vertical load. Therefore, it is a simple and efficiency way to use tire vertical load to represent the maximum of the tire adhesion capability. Based on tire friction circle concept, the tire forces are distributed in proportion to the vertical load for the UGC. It has to be mentioned that all of the chassis dynamics control block is conducted in the rapid ECU controller. In order to reduce the ECU controller's workload, a simple tire force distribution algorithm is used: in each side (left or right side), the tire forces are distributed according to the vertical load acting on tires.

An important point in this algorithm is how to calculate the tire vertical load. The tire vertical load transfer is mainly caused by the longitudinal and lateral acceleration on even road in urban environments. Therefore, the tire vertical load can be estimated by the following equations:

$$\begin{cases} F_{z11}(k+1) = -\dfrac{ma_x(k)h}{L} + \dfrac{ma_y(k)}{B}\left(\dfrac{hK_{\phi f}}{K_{\phi f}+K_{\phi r}} + \dfrac{l_r}{Lh_{rf}}\right) \\[2mm] F_{z12}(k+1) = -\dfrac{ma_x(k)h}{L} - \dfrac{ma_y(k)}{B}\left(\dfrac{hK_{\phi f}}{K_{\phi f}+K_{\phi r}} + \dfrac{l_r}{Lh_{rf}}\right) \\[2mm] F_{z21}(k+1) = -\dfrac{ma_x(k)h}{L} + \dfrac{ma_y(k)}{B}\left(\dfrac{hK_{\phi r}}{K_{\phi f}+K_{\phi r}} + \dfrac{l_f}{Lh_{rr}}\right) \\[2mm] F_{z22}(k+1) = -\dfrac{ma_x(k)h}{L} - \dfrac{ma_y(k)}{B}\left(\dfrac{hK_{\phi r}}{K_{\phi f}+K_{\phi r}} + \dfrac{l_f}{Lh_{rr}}\right), \end{cases} \tag{3.98}$$

where F_{zii} $(i=1,2)$ is the vertical load of four wheels, respectively. a_x, a_y is the longitudinal and lateral acceleration of the vehicle; h is the height of C.G; B is the track width; $K_{\phi f}$, $K_{\phi r}$ is the roll stiffness of front and rear suspension; h_{rf}, h_{rr} is the suspension roll center of front and rear suspension; and a_x, a_y is feedback by the GPS/INS system.

The tire vertical load estimation based on Equation (3.94) can be used in urban environments. However, if the UGC is working on uneven road in field environments, the vertical acceleration of the tires cannot be neglected. In this case, the estimation method based on Equation (3.94) may cause significant error. Therefore, a more direct method should be used to calculate the tire vertical load. In order to calculate the tire vertical load accurately in both urban and field environments, the linear displacement sensors are equipped on the dampers of the UGC. Therefore, the vertical load of each tire can be easily obtained based on the product of the displacement and stiffness of the dampers. Define force distribution ratio κ_l and κ_r for left and right tires:

$$\begin{cases} \kappa_l(k+1) = F_{x11}(k+1) : F_{x21}(k+1) = F_{z11}(k) : F_{z21}(k) \\ \kappa_r(k+1) = F_{x12}(k+1) : F_{x22}(k+1) = F_{z12}(k) : F_{z22}(k), \end{cases} \tag{3.99}$$

where k is the sample time step and F_{xi} ($i = 1,2$) and F_{zi} ($i=1,2$) is the traction force and vertical load of front left, front right, rear left and rear right tires, respectively.

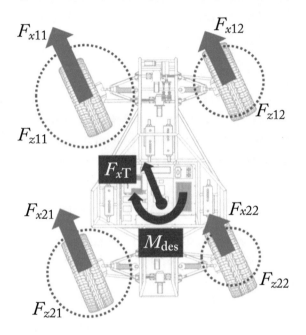

Figure 3.14: Illustration of tire forces distribution.

Figure 3.14 illustrates the tire forces distribution according to the vertical load. The target total force F_{xT} and desired active yaw moment M_{des} is determined by the upper controller. The relationship between F_{xT} and F_{xi} ($i = 1,2$) can be described as:

$$F_{xT}(k-1) = F_{x11}(k) + F_{x12}(k) + F_{x21}(k) + F_{x22}(k). \tag{3.100}$$

The steer angle of all tires is assumed equal as δ. Therefore, the relationship between M_{des} and F_{xi} ($i = 1,2$)) can be described as:

$$M_{des}(k-1) = [F_{x12}(k)\frac{B}{2}\cos\delta(k-1) + F_{x22}(k)\frac{B}{2}\cos\delta(k-1) + F_{x11}(k)l_f\sin\delta(k-1) + F_{x12}(k)l_f\sin\delta(k-1)] - [F_{x11}(k)\frac{B}{2}\cos\delta(k-1) + F_{x21}(k)\frac{B}{2}\cos\delta(k-1) + F_{x21}(k)l_r\sin\delta(k-1) + F_{x22}(k)l_r\sin\delta(k-1)]. \tag{3.101}$$

3.7.2   SLIDING MODE TIRE SLIP RATIO CONTROL

Based on Equations (3.99), (3.100), and (3.101), the desired traction force of each tire can be determined. Another major advantage of the in-wheel motor driven technique is that it is very easy to achieve optimal slip ratio control to guarantee the tire adhesion capability in various road conditions. Therefore, a sliding mode tire slip ratio controller is applied to control the tire slip ratio under the desired value. The dynamic equation of wheel can be described as:

$$I_w \dot{\omega} = i_p T_i - F_{xi} R_t - F_{Ri},$$ (3.102)

where T_i ($i = 1,2$) is the input torque of the in-wheel motors of four wheels; ω_i ($i = 1,2$) is the rotation speed of four wheels; F_{Ri} ($i = 1,$) is the resistance force of four wheels; I_w is the inertia of the wheel; R_t is the tire radius; and i_p is the transmission ratio of the planetary gear in the wheel.

The tire longitudinal slip ratio in traction condition can be described as:

$$s_i = \frac{\omega_i R_t - u}{\omega_i R_t}.$$ (3.103)

To obtain the optimal longitudinal slip ratio, the sliding mode control technique is adopted. The sliding surface of the slip ratio controller of each wheel is defined as $\lambda = s - s_d$. Differentiating the sliding surface yields:

$$\dot{\lambda} = \dot{s} - \dot{s}_d.$$ (3.104)

A continuous control law that would achieve $\dot{S}=0$ can be obtained:

$$\hat{T} = \frac{F_x R_t}{i_p} + \frac{F_r}{i_p} + \frac{I_w \dot{u} \omega}{i_p u}.$$ (3.105)

The following sliding condition is selected to avoid chattering:

$$\frac{1}{2} \frac{d}{dt} \lambda^2 \leq - \eta |\lambda|,$$ (3.106)

where η is a positive constant. The following control law with a discontinuous term can be applied to satisfy Equation (3.106):

$$T = \hat{T} - K_s \mathrm{sgn}\,(\lambda).$$ (3.107)

The control gain K_s can be calculated such that it guarantees the sliding condition. To eliminate the chattering across the surface $\lambda = 0$, it is necessary to smooth out the control discontinuity in a thin boundary layer neighboring the switching surface. Replace sgn(λ) with the continuous approximation sat(λ/Φ):

$$T = \hat{T} - K_s sat(\frac{\lambda}{\phi}) \, , \tag{3.108}$$

where *sat* is the saturation function and Φ is the boundary layer thickness which can be tuned to reduce chattering.

3.8   A NONLINEAR TIRE MODEL FOR TIRE BEHAVIOR ESTIMATION

The vehicle motion states variables feedback are very important for the dynamics control. For the UGC, most of the vehicle motion states variables directly by the GPS/INS system. In addition, the rotation speed and wheel torque can be feedback by the independent motor, and the steer torque and steer angle can be feedback by the independent steer motor. Actually, the tire behavior, especially the longitudinal and lateral tire force, is also very important for the vehicle performance estimation. However, the force sensors, which can directly measure the tire longitudinal and lateral force, are too expensive. Therefore, an analytical nonlinear tire model based on the particular condition of the UGC is proposed by the authors to provide the model for the tire behavior estimation, which will be introduced in this section.

3.8.1   THE CONTACT PRESSURE DISTRIBUTION ASSUMPTION

When the vehicle is moved on a plane, there is vertical force acting on the tires because of the gravity or vertical load transfer due to the longitudinal or lateral acceleration. The assumption of tire-ground contact pressure distribution in the contact patch is important for the modeling of analytical tire model. Among classic tire analytical models, Fiala assumes that the tire contact pressure distribution is 1D along the tire centerline. In Figure 3.15, a coordinate frame ($O'X'Y'$) is built, the length of tire contact line is $2a_l$. The Fiala model assumes the tire contact pressure distribution q_z during the contact line is described as a parabolic function:

$$\begin{cases} q_z = \frac{3}{2} \, \xi(2 - \xi) \frac{F_z}{2 \, a_l} \\ \xi = \frac{x}{a_l} \ 0 \leq \xi \leq 2 \, . \end{cases} \tag{3.109}$$

The total tire vertical force can be obtained by integrating q_z over the contact area:

$$F_z = \int_0^2 q_z d\xi. \tag{3.110}$$

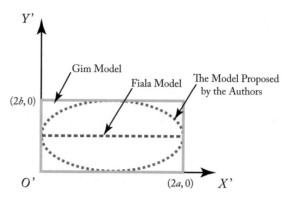

Figure 3.15: The contact patch shape assumption.

The Gim model assumes that the tire contact patch is rectangle, with length as $2a_l$ and width as $2b_w$. Along the length of tire contact patch, the contact pressure function of the tire is described as an additive of a parabolic function and a cubic function. However, it is assumed simple uniform distribution property along the tire's width direction. The following equation shows the tire-ground contact pressure distribution function of the Gim model:

$$q_z = \xi(1 - \xi)[1 + A_1(1 - A_2\xi)]\frac{F_z}{4\,a_l a_w}\,, \qquad (3.111)$$

where A_1 and A_2 are the correction factors. The tire total vertical force can be obtained by integrating the tire-ground contact pressure over the contact patch:

$$F_z = \int_{-1}^{1} d\eta \int_{0}^{2} q_z d\xi\,, \qquad (3.112)$$

where η is the non-dimensional width of the contact area:

$$\eta = \frac{y}{b_w}\ 0 \le \eta \le 2. \qquad (3.113)$$

In conclusion, the Fiala model assumes 1D tire-ground contact pressure distribution. Gim model assumes 2D tire-ground contact pressure distribution but it assumes uniform along the tire's width direction.

However, for the UGC, which is mainly used in the field environments, the side slip of tires may be too drastic to neglect the tire-ground contact pressure distribution along the tire's width direction. Therefore, it is necessary to assume the 2D tire-ground contact pressure distribution. Although Gim model assumes the 2D tire-ground contact pressure distribution, the uniform assumption along tire's width direction is not suitable for the condition of large tire side slip.

The vertical load of the tire can be considered parabolic distribution. In this tire model, both the pressure along the length direction and width direction is designed as parabolic. Obviously, the assumption of parabolic in both tire length and width directions determines that the tire-ground contact patch is an elliptic, which leads to "Elliptical Parabolic" contact pressure distribution. The tire-ground contact patch is shown in Figure 3.15 as elliptic, with semi-major axis a_l and short-half axis b_w. The value of b_w is the tire section width. The value of $2a_l$ is the contact length shown in Figure 3.16.

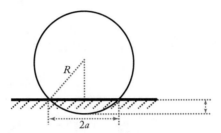

Figure 3.16: Contact length of the tire.

The tire vertical deformation can be described as:

$$\Delta = \frac{F_z}{K_t},$$

(3.114)

where K_t is the tire vertical stiffness.

Consider the kinematics relationship, and then it can be obtained:

$$a_l = \sqrt{R^2 - (R - \Delta)^2} = \sqrt{R^2 - (R^2 - \frac{F_z}{K_t})^2}.$$

(3.115)

According to the "Elliptical Parabolic" tire-ground contact pressure assumption, the distribution function can be assumed as:

$$q_z = \frac{F_z}{\pi a_l b_w}[-\frac{(\xi - 1)^2}{m^2} - \frac{(\eta - 1)^2}{n^2} + p],$$

(3.116)

where m, n, and p are the tire model parameters, which need to be determined. In order to solve these three parameters, the following relationship should be considered. First, the intersection plane between above equation and the ground plane should be a circle with radius as 1 and the circle center should be located at point (1,1). Second, the double integral of qz over the tire-ground contact elliptic patch should be equaled to the vertical force F_z:

$$F_z = \iint_C q_z \, d\xi d\eta. \tag{3.117}$$

The value of parameters m, n, and p can be solved based on the above two relationships:

$$q_z = -\frac{2F_z}{\pi a_l b_w}[(\xi - 1)^2 + (\eta - 1)^2 - 1]. \tag{3.118}$$

Equation (3.118) shows the contact pressure over the tire-ground contact patch, which can be plotted in Figure 3.17.

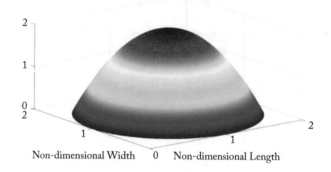

Figure 3.17: Contact pressure distribution over non-dimensional contact patch.

3.8.2 CALCULATION OF THE TIRE ACTING LONGITUDINAL AND LATERAL FORCE

The longitudinal and lateral slip of the tire occurs when the tire longitudinal and lateral force is generated. The deformation of the tire tread is shown in Figure 3.18 in the condition when the tire longitudinal and lateral force is acting. In Figure 3.18, the bold line represents the edge of the tire tread patch. Assume that the original centerline of the tire-ground contact patch is the line AD with no tire force exists. When the tire longitudinal and lateral slip occurs, the tire centerline of tire-ground contact patch changes to curve ABC.

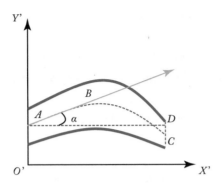

Figure 3.18: Deformation of tread when longitudinal and lateral force acts.

For each tire tread element, the deformation along the length direction is due to the tread compression to provide the tire longitudinal force, and it causes the difference between the speed of the center of the tire and the product of rotation speed and tire radius. The tire deformation along the tire width direction is due to the action of tire lateral force, which causes so-called slip angle α as. The tire longitudinal and lateral deformation can be described as:

$$\begin{cases} \Delta x = Vt \cos \alpha - \omega_i R_w t \\ \Delta y = Vt \sin \alpha. \end{cases} \tag{3.119}$$

For the convenience of describing, the tire longitudinal and lateral slip ratio can be defined as:

$$\begin{cases} S_x = \dfrac{\Delta x}{\omega R_w t} = \dfrac{Vt \cos \alpha - \omega R_w t}{\omega R_w t} \\[3mm] S_y = \dfrac{\Delta y}{\omega R_w t} = \dfrac{Vt \sin \alpha}{\omega R_w t}. \end{cases} \tag{3.120}$$

The definition domin of S_x and S_y is from the negitive infinity to the positive infinity. In adhesion area of the tire, assume the tire tread stiffness along the length and width directions is k_{tx} and k_{ty}, and then the tire shear force of a certain tire element in the adhesion aera can be described as:

$$\begin{cases} q_x = k_{tx} \Delta x = k_{tx} S_x x \\ q_y = k_{ty} \Delta y = k_{ty} S_y x. \end{cases} \tag{3.121}$$

In the tire-ground contact patch, when the tire acting force is small enough, there should be an all-adhesion area and a no-slide area. However, when the tire acting force exceeds certain value, there should be both the adhesion aera and the slide aera. When the tire acting force continues to be increased, when it exceeds the tire tread's adhesion capability, and then the whole tire-ground contact area will start to slide.

Assume there is no slide in the tire-ground contact patch, and then the tire longitudinal and lateral force can be obtained:

$$\begin{cases} F_x = \iint_D q_x \, dxdy = a_l^2 b_w \pi k_{tx} S_x = K_X S_x \\ F_y = \iint_D q_y \, dxdy = a_l^2 b_w \pi k_{ty} S_y = K_Y S_y \, . \end{cases} \tag{3.122}$$

As the increase of acting force, for each tire tread element, the slide phenomenon will occur when the tire resultant shear force equals to the product of tire-ground adhesion coefficent and the tire vertical force:

$$\sqrt{q_x^2 + q_y^2} = \mu q_z. \tag{3.123}$$

Substitute Equations (3.121) and (3.122) into (3.123):

$$\sqrt{(k_{tx} + S_x)^2 + (k_{tx} + S_y)^2} \cdot x = \mu q_z, \tag{3.124}$$

where x is the point of the tread where the slide starts, the tire contact patch before point x remains adhesion situation, and the tire contact patch after point x begins to slide.

When there are both adhesion and slide areas in the contact patch, the tire resultant shear force is obtained by:

$$F = \iint_D \sqrt{q_x^2 + q_y^2} dxdy + \iint_E \mu q_z dxdy. \tag{3.125}$$

Actually, there are two parts of double integration caculation: integration domains D and E, which is the adhesion area and slide area, respectively. The normailized tire longitudinal slip ratio, tire lateral slip ratio, and tire conbined slip ratio are defined as:

$$\begin{cases} \Phi_x = \dfrac{K_x S_x}{\mu F_z} \\[2ex] \Phi_y = \dfrac{K_y S_y}{\mu F_z} \\[2ex] \Phi = \dfrac{\sqrt{(K_x S_x)^2 + (K_y S_y)^2}}{\mu F_z} \end{cases} . \tag{3.126}$$

After the manipulation, the final expression of the tire resultant force can be obtained:

$$F = \mu F_z \cdot (-\frac{\Phi^4}{256} + \frac{\Phi^3}{16} - \frac{3\Phi^2}{8} + \Phi). \tag{3.127}$$

When the whole contact patch of the tire begins to slide, the tire resultant force can be calculated:

$$F = \mu F_z. \tag{3.128}$$

Based on this tire analytical model, the tire longitudinal and lateral force can be estimated online if the corresponding motion variables can be collected, such as the tire rotation speed, tire longitudinal slip ratio, and the tire lateral slip angle (caculated by the vehicle yaw velocity and the vehicle sideslip angle).

CHAPTER 4

Performance Test and Evaluation of the Unmanned Ground Carrier

In this chapter, the results of the experiments of the UGC will be shown and discussed to evaluate the performance of the UGC. The experiments include a variety of tests to test different kinds of performance of the UGC, such as the endurance performance, acceleration performance, performance of the chassis dynamics controller, and the performance to send out or take back the rotorcrafts/robots. In most of the experiments, both the experiment results and the experiment scenes are given to fully discuss the results.

The equipped sensors and the data acquisition system of the UGC need to be mentioned. In order to collect the vehicle position and motion data information, a GPS/INS system is equipped on the UGC. The motor torque and wheel rotation speed is feedback by the independent motors. Similarly, the steer angle of each wheel is collected by each steer servomotor. The driven motor's information, such as the current and the voltage, can be collected by the motor controller. The vertical displacement of the wheel can be collected by the damper linear displacement sensor. The state of the battery pack and each battery can be collected by the BMS. These sensors and the data acquisition system enables us to make a comprehensive evaluation of the UGC's performance.

4.1 BASIC PERFORMANCE TEST AND ENDURANCE PERFORMANCE EVALUATION IN FIELD ENVIRONMENTS

4.1.1 A BASIC TEST FOR THE UGC—THE MISSION TO FIND A TARGET IN THE NARROW SPACE

At first, the basic performance of the UGC should be tested, that is, carrying the onboard robot to conduct the mission collaboratively. Actually, the basic performance test can carry out the reliability of all the subsystems and the components of the whole UGC, including the basic performance of the command station, mother vehicle, and the carry-on robot. In order to test the ability of the UGC in the complex field environment. The experiment site is selected as a very muddy environment. The mission can be summarized as: the human operator is in a nearby building to control the command station. The UGC is sent out by the human operator to find a target in the narrow space. In addition, the mother vehicle should sent out the robot to conduct the mission and then

(a) The UGC is operating in the muddy road

(b) The human operator is controlling the command station

(c) The robot is sent out to find the target in the grass

(d) The UGC takes the robot back to the base

Figure 4.1: The experiments to conduct the mission of target detecting in the grass.

take it back. This mission can be considered as a basic test for the UGC. The pictures during the experiments are shown in Figure 4.1.

The pictures that were taken during the experiments are shown in Figure 4.1. According to Figure 4.1(b), the human operator is sitting in a nearby building to manipulate the command station to control the UGC, as well as watching the monitor to observe the environments. The UGC is controlled by the human operator to travel through a muddy area to achieve the target area as Figure 4.1(a) shows. The distance of the muddy area is about 1 km. Figure 4.1(c) shows that after the UGC achieves the target area. The human operator controls the door in the front of the vehicle to be open to send the robot out, and the robot is moving into the grass to find the target. After the target is found, the robot goes back to the mother vehicle by climbing the slide. Finally, as Figure 4.1(d) shows, the mother vehicle takes the robot back to the human operator.

This simple experiment clearly shows the advantage of the UGC. At first, if the robot itself is sent out to cross the muddy road, it will be very difficult for it to finish the mission. The small robot is not able to negotiate the muddy road. In addition, the locomotion speed of the robot is too low, therefore, it will cost lots of time to finish the mission. However, the mother vehicle can negotiate the muddy road easily based on its good driving performance and high driving power. Second, if the mother vehicle itself is sent out to conduct the mission, it will be very difficult for it to find the target in such a narrow space (grass). The size of the mother vehicle is too large. Therefore, the advantages of the mother vehicle and the small robot have been combined perfectly to conduct this mission. In addition, the basic performance and reliability of all the components of the UGC were proved by this experiment.

4.1.2 THE ENDURANCE EXPERIMENTS IN THE FIELD ENVIRONMENTS

After the experiment of testing the basic performance, the endurance performance of the UGC will be tested. The ability to operate mutely is very important for a military vehicle, especially because it enables the vehicle to operate in the target area and not be detected by the enemy. It is very important to improve the invisibility and the ability to survive for a military vehicle. Apparently, this is a great advantage of the pure electric powered technique. Based on pure electric powered technique, the UGC is able to operate in mute mode.

However, the endurance performance may be a big problem for the pure electric-powered vehicle because of the limit of the battery pack capacity. Although the UGVs modified based on ICE vehicles may have better endurance performance, however, the other performance of these modified UGVs will be degenerated, which has been described in Chapter 1. Therefore, how to improve the endurance performance of the pure electric powered vehicle is always an important issue to be dealt with for engineers and researchers. In addition, the military vehicles are mainly used for field environments. In field environments, the vehicle may need much more power to overcome the resistance force caused by the field road. At some time, the vehicle needs to climb some slopes or negotiate some other complex terrains in the field environment, which may cost even more power. Therefore, it has been widely accepted that the military vehicle, which is always used in field environments, needs much more battery capacity and much better endurance performance.

As introduced in Chapter 3, in order to overcome this problem and to improve the endurance performance of the UGC, the battery pack on the UGC is designed as a high-power density battery pack with advanced BMS technique, which enables the UGC to work in the field environments for about 100–200 km. Especially, the discharging rate of the battery cell is designed to be 10C to significantly improve the power density of the battery pack. In order to test the endurance performance of the UGC and the reliability of the battery pack, the endurance tests in field environments are conducted first. In addition, the meaning of the endurance tests is not only for the endurance performance evaluation, but also an opportunity to test the performance of all other performances.

(a) Incline climbing area

(b) Grass and soft sand area in field environments

(c) Muddy road area in field environments

Figure 4.2: Experiment scenarios in field environments.

Table.4.1: Record of the endurance experiments	
Parameters	**Values**
Endurance Time	6 h
Endurance Distance	143 km
State of Charge (SOC) before Experiments	80%
SOC After Experiments	27%
Highest Speed	37 km/h
Lowest Speed	17 km/h
Maximum Current	507 A
Minimum Current	121 A
Temperature of Cooling Water before Experiments	25°C
Maximum Temperature of Cooling Water	70°C

Figure 4.2 shows some pictures during the endurance experiments in field environments. The experiments are conducted in a military test field. As Figure 4.2(a) shows, the selected experiment site includes three typical areas: incline and sand road areas; grass and soft sand areas; and muddy road areas. These areas are very typical in a military test field. Before the experiments, the battery pack should be fully charged, which is necessary to conduct the endurance performance experiment. During the experiment, the human operator is in the building to control the UGC through the command station. The experiment mission can be summarized as: the human operator controls the UGC to operate in these areas to fully discharge the battery pack to test the endurance performance. In each area, the human operator should control the UGC to travel a circle, and then control it to transfer to the next area. In other words, the UGC should be controlled to continuously operate in the grass, soft sand, muddy road, incline, and sand road to test the endurance performance.

Figures 4.2(a–c) show the UGC is operating in these three areas. Generally, these three areas represent different typical working condition of the UGC. For example, in grass and soft sand areas, the speed of the UGC is relatively high since the road in this area is relatively flat. Therefore, this area represents the general working condition of the UGC in the field environments. In muddy road area, the speed of the UGC is the lowest due to the large resistance force of the muddy road. Therefore, this area represents the worst working condition of the UGC in the field environments. It can be seen in Figure 4.2 that the muddy road is very difficult to travel. In incline and sand road areas, shown in Figure 4.2(c), the UGC should be controlled to operate with the highest power to climb the incline and slope, which is an important test during the endurance experiments. The experiments will be continued until the battery is fully discharged.

In order to discuss the endurance performance, the parameters related to the endurance performance are recorded, which are shown in Table 4.1. The battery pack is fully charged before the endurance experiments. It can be seen that the SOC before the experiments is 80%. If the SOC of the battery cell is higher than 80%, the battery cell will be easy to be damaged. Therefore, the SOC maximum limit is designed as 80%. The SOC value continues to decrease during the endurance experiment. After the experiments, the SOC value becomes 27%, which means the battery pack is almost fully discharged. Similarly, if the SOC value of the battery cell is too low, the battery cell will also be easy to be damaged.

Therefore, the minimum limit is designed as 30%. According to the strategy of the BMS, if the SOC of the battery pack is lower than 30%, the battery pack will be defined as "fully discharged." Therefore, the BMS cuts the power when the SOC value is 27%.

The experiments continuously last for 6 h, and several human operators control the UGC turn by turn. The traveled distance during the experiments is about 143 km, which is collected by the GPS system. Apparently, the designed goal of the endurance distance (100–200 km) is almost achieved. The highest speed of the UGC during the experiment is 37 km/h, which occurs in the grass and soft sand area. The lowest speed of the UGC during the experiment is 17 km/h, which occurs in the muddy road area. The maximum current of the battery pack is 507 A, which occur during the UGC climbing on the slope. As above mentioned, the maximum torque is needed when the UGC is climbing on the slope. The value of 507 A is almost the maximum current capacity of the whole battery pack, which means the peak power is outputted during the slope climbing. In addition, the performance of the whole cooling system is very important to evaluate the endurance performance, which can be represented by the temperature of the cooling water. This can be recorded by the temperature sensor equipped in the cooling system. The maximum temperature of the cooling water is 70°C, which is a desirable value. This proves the efficiency of the whole cooling system. Generally, the experiments results successfully prove the endurance performance and the reliability of the UGC, as well as the reliability of all the components.

4.1.3   THE TEST OF THE MECHANISMS TO SEND OUT AND TAKE BACK THE ROBOT/ROTORCRAFTS

In addition, the mechanism to send out and take back the robot/rotorcrafts should be tested individually. The robot and the rotorcrafts carried on the UGC are not developed by the authors, and only the communication systems are modified to obtain collaborative control through command station. Therefore, the experiments to evaluate the performance of the robot and rotorcrafts are not conducted, and only the experiments to evaluate the performance of the sending out and taking back mechanisms are conducted. Figure 4.3 shows the experiments to send out the robot. It takes 5 s to send out the robot. Figure 4.3(a) shows the door is closed at 0 s. The servomotors control the door to be open and the slide to be sent out, which takes about 2 s. The robot climbs down through the slide, and the whole procedure takes about 5 s. Figure 4.4 shows the experiment to take back the robot, and the whole procedure takes about 6 s. Figure 4.5 shows the experiments to take back the rotorcraft. As introduced above, there are electromagnets equipped on the platform to stable the rotorcraft. The electromagnets can be controlled to release the rotorcraft by the switch on the command station. After the rotorcraft successfully takes off, the experiment to take back the rotorcraft will start. It takes about 6 s to land the rotorcraft in the experiment. After the rotorcraft is landed on the platform, the electromagnets can be controlled to be turned on to stable the rotorcraft by the switch on the command station. The experiments successfully prove the effect of the mechanism and the control system.

(a) 0 s

(b) 2 s

(c) 4 s

(d) 5 s

Figure 4.3: The experiment to send out the robot.

(a) 1 s

(b) 3 s

(c) 5 s

(d) 6 s

Figure 4.4: The experiment to take back the robot.

(a) 0 s

(b) 2 s

(c) 3 s   (d) 5 s

Figure 4.5: The experiment to take back the rotorcraft.

4.2 ACCELERATION AND TIRE FORCE DISTRIBUTION PERFORMANCE EVALUATION

As widely known, the acceleration performance, especially in field environments, is very important for a military vehicle. Generally, it is also an important index to evaluate the mobility of the military vehicle. As mentioned above, the drive-by-wire technique can be used to significantly improve the acceleration and driving performance of the vehicle. With the in-wheel motor driven technique, the good acceleration and driving performance is an important feature of the UGC. At first, the maximum torque capability and the transient response performance of the electric motor is much better than that of the traditional engine. In addition, the force distribution strategy based on the vertical load (FDSVL) can further improve the acceleration performance of the UGC.

4.2.1 THE ACCELERATION EXPERIMENTS ON PAVEMENT ROAD

In order to evaluate the acceleration performance of the UGC, the acceleration experiments on both pavement and field roads are conducted. To verify the efficiency of the proposed FDSVL, both the experiments with and without FDSVL are conducted. In each test, the UGC is controlled to accelerate in a straightline with the maximum power, and the value of speed, tire vertical load, motor torque, and tire slip ratio are recorded. The vehicle speed is collected by the GPS system. The tire vertical load is collected by the displacement sensor of the damper. The motor torque is feedback by the motor controller. The tire slip ratio is calculated based on the wheel rotation speed and the vehicle speed.

The recorded experiment results ona paved road are shown in Figure 4.6. The road friction coefficient is about 0.5–0.6, since the pavement road is covered with sand. The vehicle speed comparison between the results with and without FDSVL is shown in Figure 4.6(a). The FDSVL greatly improves the acceleration performance of the vehicle. The acceleration time from zero to

100 km/h is about 7.5 s under FDSVL, and it is about 8.5 s without FDSVL. The acceleration time from 0 to 100 km/h is an important index to evaluate a military vehicle's mobility. This index of the UGC is much higher than that of other level military vehicles.

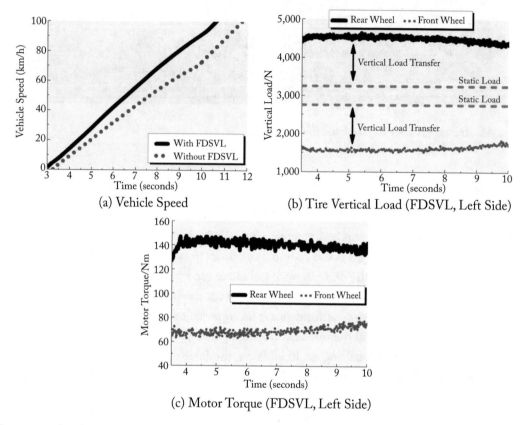

(a) Vehicle Speed   (b) Tire Vertical Load (FDSVL, Left Side)

(c) Motor Torque (FDSVL, Left Side)

Figure 4.6: Acceleration experiments results in pavement road.

The data of tire vertical load and motor torque during the experiment under FDSVL is shown in Figures 4.6(b, c). In order to give a clear observation, only the results of left-side wheels are given. It can be seen that great vertical load transfer occurs due to the large acceleration. The static load of the front left wheel is about 2,800 N, and the dynamic load of the front left wheel during acceleration is about 1,500 N. The static load of the left wheel is about 3,200 N, and the dynamic load of the rear left wheel during acceleration is about 4,500 N. It can be concluded that the vertical load of rear left wheel greatly increases from the static vertical load, and the vertical load of front left wheel greatly decreases. Figure 4.6(c) shows the motor torque value of the left front and rear wheels. It can be seen that the motor torque is distributed according to the tire vertical load value, which proves the effect of the FDSVL.

4.2.2   THE ACCELERATION EXPERIMENTS ON OFF ROAD

(a) 0.00 s

(b) 0.03 s

(c) 0.05 s

(d) 0.06 s

Figure 4.7: Acceleration experiments in field road.

The recorded experiment results on field off-road conditions are shown in Figure 4.8. The experiments are conducted in an open ground in field muddy road, and the experiment scenarios are
shown in Figure 4.7. The comparison between the results with and without FDSVL is shown in
Figure 4.8(a). It can be easily seen that the FDSVL significantly improves the acceleration performance. The UGC with FDSVL reaches 30 km/h at 5 s, and the UGC without FDSVL only
reaches 22.5 km/h at 7 s. In addition, the optimal traction force distribution also increases the
highest speed in off road. The highest speed of the vehicle without FDSVL is only 22.5 km/h and
the highest speed of the vehicle with FDSVL is 30 km/h. It is because the FDSVL significantly
improve the driving efficiency of the UGC. Figure 4.8(b) shows the tire slip ratio without FDSVL
or tire slip ratio controller. It can be seen that the tire slip ratio is very large, which already reaches
1 at some time. The average value of the tire slip ratio is about 0.7. The tire slip ratio is 1 means
that the rotation speed of the wheel is much higher than the vehicle speed, which indicates the
extreme slip and skid phenomenon occurs in off road. The tire adhesion capability will be greatly
degenerated if large tire skid and slip occurs. Therefore, the acceleration and driving performance of
the UGC is greatly degenerated by the large tire skid and slip ratio. Figure 4.8(c) shows the motor
torque distribution under FDSVL and tire slip ratio controller. The motor torque is distributed

according to the vertical load and adjusted to maintain a desirable tire slip ratio level. Figure 4.8(d) shows the tire slip ratio under FDSVL and tire slip ratio controller. It can be seen that the tire slip ratio has been greatly reduced by FDSVL and tire slip ratio controller. The average of the tire slip ratio is only about 0.3. In conclusion, when the vehicle is operating in the field off-road condition, the acceleration and driving performance will be greatly degenerated due to the large tire slip ratio and the decrease of tire adhesion capability. The tire slip ratio controller proposed in this book can successfully reduce the tire slip ratio and improves the driving performance off road.

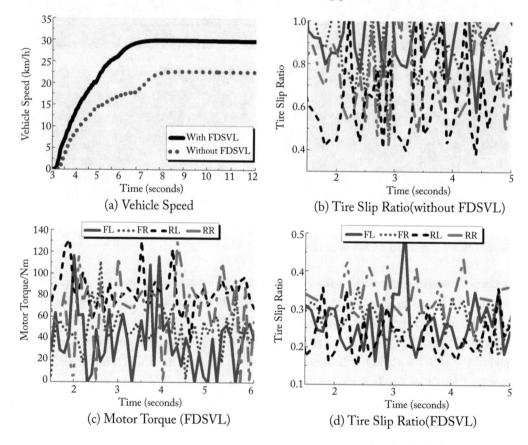

Figure 4.8: Acceleration experiments results in field road.

4.3 CHASSIS YAW MOMENT CONTROL PERFORMANCE EVALUATION

4.3.1 YAW MOMENT CONTROL EXPERIMENT IN ASM ON PAVEMENT ROAD

As described above, the yaw moment controller aims at improving the stability in the high-speed maneuver in ASM, and improving the path-tracking performance in the high-speed maneuver in DMSM. At first, the experiments are conducted to verify the pole assignment yaw moment controller in high speed maneuver in ASM. In the experiments, the UGC is still remotely controlled by the human operator through the command station. The step steer signal input test is very common used in vehicle handling performance test. Therefore, the step steer signal input tests are conducted to make a comparison of the vehicle handling performance with or without control. Before the experiments, the following points should be noticed.

1. As described above, the yaw moment controller is designed based on the pole assignment technique, which allows the designers to select the target poles locations to adjust the desired handling performance of the UGC. Therefore, in the experiments, two cases of target closed-looped poles locations are selected to carry out this benefit of the controller.

2. In order to show the performance of the yaw moment controller more comprehensively, both the experiments in pavement road condition and off-road condition are conducted.

3. The robustness performance is very important for the chassis dynamics controller. In order to test the robustness performance of the controller, the UGC in pavement road condition experiments is with no load to obtain minimum vehicle mass, and the UGC in off-road condition experiments is fully loaded to obtain maximum vehicle mass.

In each of the experiment, the target vehicle speed of the UGC is remotely controlled by the human operator based on the drive lever on the command station, and the target step steer signal is inputted by a program by ECU controller.

The recorded experiment results in pavement road condition are shown in Figure 4.9. Three different cases of tests are conducted. The first category of the test is the UGC without the yaw moment control. The second category of the test is the UGC under yaw moment control, and the target closed-looped poles locations are defined as a value is 0.8 and b value is 6. The third category of the test is still the UGC under yaw moment control, but the target closed-looped poles locations are defined as a value is 0.8 and b value is 4. According to the definition of parameters a and b, it can be known that the static margin in case 2 is higher than that in case 3. Therefore, the handling stability performance of the vehicle in case 2 is better than that of the vehicle in case 3, and the mobility of the vehicle in case 3 is better than that of the vehicle in case 2. As mentioned above, an important advantage of the robust pole assignment yaw moment controller is that it enables the designers to adjust the target closed-looped poles locations to change the desired vehicle handling performance based on different requirements. Apparently, these experiments show this advantage and benefit.

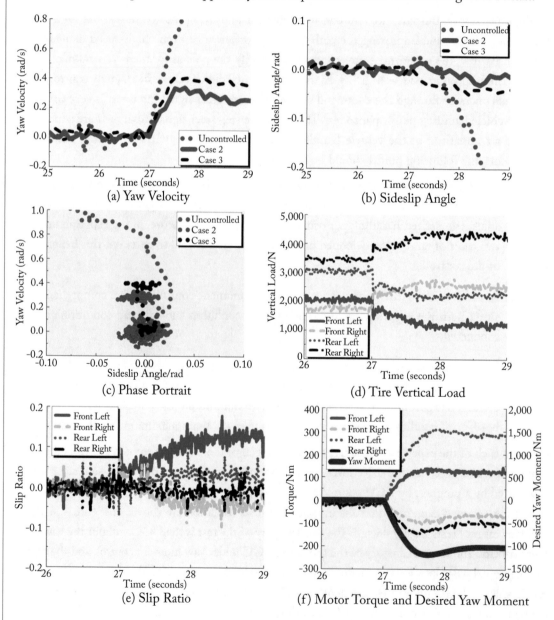

Figure 4.9: Experiment results on pavement road condition in ASM.

The target vehicle speed in the experiments is 80 km/h. The step steer signal is inputted at 27 s. Figures 4.9(a, b) show the comparison of the value of vehicle yaw velocity and vehicle sideslip angle. When the step steer angle signal is inputted, the vehicle yaw velocity and vehicle sideslip angle value in uncontrolled case increases rapidly, which is divergent. It can be concluded that the uncontrolled UGC completely loses the stability and begins to spin. At 27.5 s, the vehicle yaw velocity value reaches as high as 0.8 rad/s. At 28.5 s, the vehicle sideslip angle value reaches as high as -0.2 rad. As shown above, the mother vehicle of the UGC is inherent unstable and oversteer. Therefore, it loses the stability in such a high vehicle speed steering maneuver. In contrast, the UGC in case 2 behaves very well after the step steer angle signal is inputted. The value of yaw velocity and the sideslip angle value stabilizes at 0.3 rad/s and -0.02 rad. It can be known that the pole assignment yaw moment chassis controller successfully keeps the handling stability of the UGC. In addition, the UGC in case 3 is also successfully stabilized by the yaw moment chassis controller. The yaw velocity value stabilizes at 0.37 rad/s. The yaw velocity value in case 3 is a little higher than that of the UGC in case 2, which accords to the predefined control goal. As discussed above, the static margin of the target closed-looped poles locations in case 2 is higher than case 3. Therefore, the handling stability in case 2 is better, and the mobility in case 3 is better, which has also been successfully proved by the above experiments results. In order to give a clear observation, the phase portrait plotted by yaw velocity value against sideslip angle value is shown in Figure 4.9(c). It can be easily seen that the states variables of the UGC without yaw moment control exceeds the stable safe boundary and becomes divergent.

Figure 4.9(d) shows the tire vertical load value in case 2. The vertical load of each tire is calculated by multiplying the value of the displacement of the damper and the damper stiffness value. Before 27 s, the UGC is working in straightline condition. The tire vertical load value between left and right tires is different. Since a left-hand turn steer angle signal is inputted at 27 s, the tire vertical load transfers from the left tires to the right tires due to the occur of the lateral acceleration. The tire vertical load of the rear right tire is the largest value, which is 4,000 N. The tire vertical load of the front left tire is the smallest value, which is 1,000 N. Figures 4.9(e, f) show the value of tire slip ratio, motor torque acting on wheels and the desired yaw moment value in case 2. In this left-hand turn, a negative yaw moment is inputted to keep the UGC from losing the handling stability. The peak value of the desired yaw moment is -1,200 Nm, which occurs at about 27.5 s. Therefore, the motor torque provided by the left wheel motors is positive and the motor torque provided by the right wheel motors is negative. The motor torque acting on the rear left tire is the largest value, which is about 300 Nm. The torque acting on the wheel is calculated based on the torque value feedback and the gear ratio of the planetary gearbox. Since the experiments are on the pavement road, the slip ratio value of the tires is small. The slip ratio value of the front left tire is largest due to its smallest vertical load. Since the motors of left tires provide positive motor torque and motors of right tires provide negative motor torque, the slip ratio value of left tires is positive and that of right tires is negative.

4.3.2   YAW MOMENT CONTROL EXPERIMENT IN ASM ON OFF ROAD

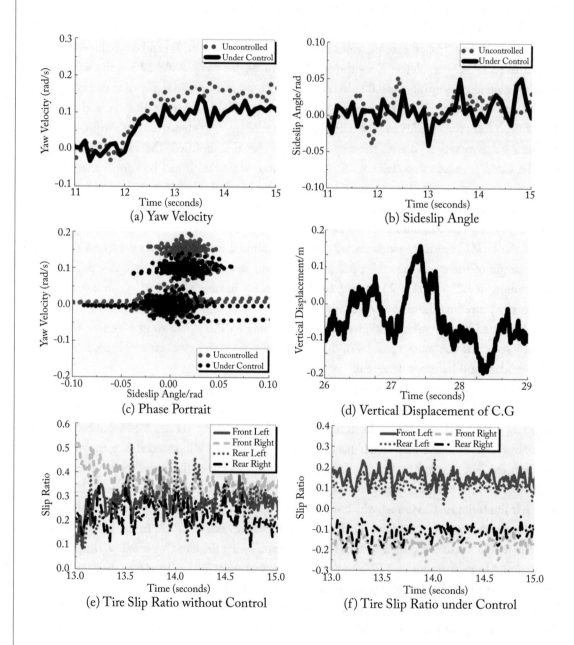

(a) Yaw Velocity

(b) Sideslip Angle

(c) Phase Portrait

(d) Vertical Displacement of C.G

(e) Tire Slip Ratio without Control

(f) Tire Slip Ratio under Control

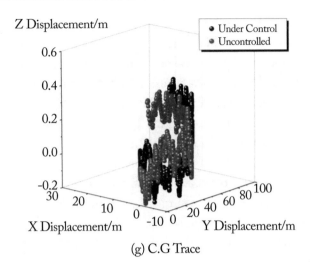

(g) C.G Trace

Figure 4.10: Experiment results in off-road condition in ASM.

In the experiments on off-road conditions, the target vehicle speed is selected as 30 km/h and the step steer angle signal is inputted at 12 s. The values of parameters a and b are selected as 0.7 and 5. The main purpose of these off-road condition experiments is to carry out the performance of yaw moment controller in off-road conditions, as well as the performance of the motor forces distribution strategy and the slip ratio controller. Therefore, only one case of target closed-looped poles locations is selected.

Figures 4.10(a, b) show the comparison of the vehicle yaw velocity and vehicle sideslip angle. Since these experiments are conducted on off-road conditions, the noise of these collected data is larger than those in the experiments on the pavement road. It can be seen that the uncontrolled UGC doesn't lose the stability, since the operating vehicle speed is relatively low (30 km/h). The yaw velocity value of the uncontrolled UGC stabilizes at the 0.15 rad/s. However, the yaw moment chassis controller reduces the response value of the vehicle yaw velocity, which indicates that the stability of the UGC has been improved. In addition, the transient response speed is improved and the transient response period is reduced. The value of sideslip angle of uncontrolled and under controlled UGC is relative small. According to Figure 4.10(c), the range of the states variables of under controlled UGC is smaller than that of the uncontrolled UGC, which also represents the better handling stability of the UGC. The vertical displacement value of the C.G location of under controlled UGC is shown in Figure 4.10(d), which is collected by the GPS system. Based on Figure 4.10(d), the situation of road condition can be known. The maximum of the vertical displacement of C.G location is 0.3 m, and the minimum value is about -0.2 m. The vertical displacement of C.G location fluctuates greatly due to the uneven road.

The slip ratio value of the uncontrolled UGC is shown in Figure 4.10(e). In off-road conditions, the vertical load value of each tire changes largely, some tires even lose the contact with the ground plane. Therefore, the tire slip ratio value of the tires is relative large and changes from 0.1 to 0.5 during the whole experiment, which means the adhesion situation of tire is very bad. Figure 4.10(f) shows the tire slip ratio value of the under controlled UGC. The tire slip ratio decreases because of the involvement of tire slip ratio control. The value of tire slip ratio range of the front left and the rear left tires is from about 0.05 to about 0.2, and the value range of the front right and the rear right tires is from about -0.05 to about -0.25. The desired tire slip ratio value of the slip ratio controller is selected as 0.15. Since the motors of left wheels provide the positive torque, and the motors of right wheels provide the negative torque, the slip ratio value of the left tires is positive and the slip ratio value of the right tires is negative. The decrease of the value of tire slip ratio improves the tire's adhesion capability, and consequently enhances the handling performance of the UGC.

4.4 TIRE SLIP RATIO CONTROL AND ZRSM PERFORMANCE EVALUATION

The performance of the tire slip ratio controller is verified by an individual experiment, which is shown in Figure 4.11. The UGC is in muddy road, and the drive lever is fully operated to give the full motor torque. To compare the wheel behavior with and without slip ratio control, the slip ratio controller is applied on rear right wheel, and not applied on front right wheel. Figure 4.12 shows eight segments in 1 s. It can be seen that the front right wheel greatly spins during the experiment, since the road friction coefficient of the muddy road is very low and the slip ratio controller is not applied for the front right wheel. A red point is marked on one of the six bolts on the rear right wheel. It can be seen that the rear right wheel doesn't spin because of the application of tire slip ratio controller. It successfully adheres on the ground and gradually puts the UGC out of the muddy road.

In addition, the performance in ZRSM will also be evaluated. In ZRSM, all the wheels are placed in the positions as Figure 4.12 and Figure 4.13 show, which provides the possibility for the UGC to achieve pivot steer to greatly improve the mobility in narrow spaces. Compared to skid steer, which is a common technique for six-wheeled military UGV, the ZRSM can significantly reduce the tire slip and skid to improve the performance. The ZRSM experiments in both field (sand road) and urban (pavement road) environments are conducted, which are shown in Figure 4.12 and 4.13. The time cost to finish one circle on sand road is only 1.8 s, and the time cost to finish one circle on pavement road is only 2.5 s. This is much faster than other military UGVs, which can also achieve pivot steer.

Figure 4.11: Performance of the tire slip ratio controller.

Figure 4.12: The experiments in ZRSM in field environments.

Figure 4.13: The experiments in ZRSM in urban environments.

References

[1] M. Montemerlo, J. Becker, S. Bhat, H. Dahlkamp, D. Dolgov, S. Ettinger., D. Haehnel, T. Hilden, G. Hoffmann, B. Huhnke, D. Johnston, S. Klumpp, D. Langer, A. Levandowski, J. Levinson, J. Marcil., D. Orenstein, J. Paefgen, I. Penny, A. Petrovskaya, M. Pflueger, G. Stanek, D. Stavens, A. Vogt, and S. Thrum. 2008. Junior: The Stanford entry in the urban challenge. *Journal of Field Robotics*, 25(9), pp. 569–597. DOI: 10.1002/rob.20258. 2, 4, 10

[2] http://news.china.com/news100/11038989/20170905/31291136.html. 5

[3] https://baijiahao.baidu.com/s?id=1570604191732340&wfr=spider&for=pc. 5, 6

[4] http://tech.sina.com.cn/it/2017-04-13/doc-ifyeimqc3234830.shtml. 6

[5] http://www.bilibili.com/video/av4480358/. 7

[6] http://v.youku.com/v_show/id_XNjcyOTAwOTk2.html. 8

[7] http://military.china.com.cn/2015-07/10/content_36028781_6.htm. 9

[8] http://www.jhbaobei.com/junyong/2017-03-12/1270.html. 9, 10

[9] J. Loenard, J. How, S. Teller, M. Berger, S. Campbell, G. Fiore, L. Fletcher, E. Frazzoli, A. Huang, S. Karaman, O. Koch, Y. Kuwata, D. Moore, E. Olson, S. Peters, J. Teo, R. Truax M. Walter, D. Barrett, A. Epstein, K. Maheloni, K. Moyer, T. Jones, R. Buckley, M. Antone, R. Galejs, S. Krishnamurthy, and J. Williams. 2008. A perception-driven autonomous urban vehicle. *Journal of Field Robotics*, 25(10), pp.,727–774. 10

[10] C. Urmson, J. Anhalt, H. Bae, J. A. (Drew) Bagnell, C. R. Baker, R. E. Bittner, T. Brown, M. N. Clark, M. Darms, D. Demitrish, J. M. Dolan, D. Duggins, D. Ferguson, T. Galatali, C. M. Geyer, M. Gittleman, S. Harbaugh, M. Hebert, T. Howard, S. Kolski, M. Likhachev, B. Litkouhi, A. Kelly, M. McNaughton, N. Miller, J. Nickolaou, K. Peterson, B. Pilnick, R. Rajkumar, P. Rybski, V. Sadekar, B. Salesky, Y.-W. Seo, S. Singh, J. M. Snider, J. C. Struble, A. (Tony) Stentz, M. Taylor, W. (Red) L. Whittaker, Z. Wolkowicki, W. Zhang, and J. Ziglar. 2008. Autonomous driving in urban environments: Boss and the urban challenge. *Journal of Field Robotics*, 25(8), pp. 425–466. DOI: 10.1002/rob.20255.

[11] S. Kammel, J. Ziegler, B.Pitzer, M. Werling, T. Gindele, D. Jagzent, J. Schroder, M. Thuy, M. Goebl, F. von Hundelshausen, O. Pink, C. Frese, and C. Stiller. 2008. Team AnnieWAY's autonomous system for the 2007 DARPA urban challenge. *Journal of Field Robotics*, 25(9), pp. 615–639. DOI: 10.1002/rob.20252. 11

[12]   https://en.wikipedia.org/wiki/Unmanned_ground_vehicle. 14, 18

[13]   J. Ni and J. Hu. 2016. Dynamic modeling and experimental validation of skid steered vehicle in the pivotal steer condition. *Proceedings Institution Mechanical Engineering, Part D: Journal Automobile Engineering*, 231(2), pp. 1–16. 15, 16, 18

[14]   Z. Zhang, X. Zhang, and H. Pan, W. Salman, Y. Rasim, X. Liu, C. Wang, Y. Yang, and X, Li. 2017. A novel steering systems for a space-saving 4WS4WD electric vehicle: Design, modeling and road tests. *IEEE Transactions Intelligent Transportation Systems*, 18(1), pp. 114–127. DOI:10.1109/TITS.2016.2561626. 21, 22

[15]   Y. Hori. 2004. Future vehicle driven by electricity and control-research on four-wheel-motored UOT March II. *IEEE Transactions Vehicular Technology*, 51(5), pp. 954–962. DOI: 10.1109/TIE.2004.834944. 21, 22

[16]   K. Nam, H. Fujimoto, and Y. Hori. 2014. Advanced motion control of electric vehicles based on robust lateral tire force control via active front steering. *IEEE Transactions Vehicular Technology*, 19(1), pp. 289–299. DOI: 10.1109/TMECH.2012.2233210. 21, 22

[17]   R. Wang, Y. Chen, D. Feng, X. Huang, and J. Wang. 2011. Development and performance characterization of an electric ground vehicle with independently actuated in-wheel motors. *Journal of Power Sources*, 196, pp. 3962-3971. DOI: 10.1016/j.jpowsour.2010.11.160. 21, 22

[18]   http://www.bit.edu.cn/xww/lgxb21/128711.htm. 25

[19]   http://www.miit.gov.cn/n1146290/n1146402/n1146445/c5775030/content.html. 25

[20]   http://news.sina.com.cn/o/2017-08-29/doc-ifykiurx2601953.shtml. 25

[21]   J. Ni and J. Hu. Development of a mother-child and X-by-wire unmanned ground vehicle: Unmanned ground carrier. *Journal of Field Robotics* (Under Review). 27, 29, 33, 35

[22]   J. Ni, J. Hu, and C. Xiang. An AWID and AWIS X-by-wire UGV: Design, dynamics control, and experiment. *IEEE Transactions Intelligent Transportation Systems* (Under Review). 29, 35, 69

[23]   J. Ni, J. Hu, and C. Xiang. Robust control in diagonal move steer mode and experiment on an X-by-wire UGV. *IEEE/ASME Transactions Mechatronics* (Under Review). 29, 35, 77

[24]   I. Genya, K. Iagnemma, J. Overholt, and G. Hudas. 2015. Design, development, and mobility evaluation of an omnidirectional mobile robot for rough terrain. *Journal Field Robotics*, 32(6), pp. 880–896. DOI: 10.1002/rob.21557. 41, 42

[25]   C. C. Chan. 2008. The state of the art of electric and hybrid vehicles. *IEEE Transactions Industrial Electronics*, 55(6), pp. 2237–2245. 53

[26]   L. D. Novellis, A. Sorniotti, and P. Gruber. 2014. Wheel torque distribution criteria for electric vehicles with torque-vectoring differentials. *IEEE Transactions Vehicular Technology*, 63(4), pp. 1593–1602. DOI: 10.1109/TVT.2013.2289371. 53

[27]   G. A. Magallan, C. H. Angelo, and G. O. Garcia. 2011. Maximization of the traction forces in a 2WD electric vehicle. *IEEE Transactions Vehicular Technology*, 60(2), pp. 369–380. DOI: 10.1109/TVT.2010.2091659. 53

[28]   S. Khosravani, A. Kasaiezadeh, A. Khajepour, B. Fidan, S-K Chen, and B. Litkouhi. 2015. Torque vectoring based vehicle control robust to driver uncertainties. *IEEE Transactions Vehicular Technology*, 64(8), pp. 3359–3367. DOI: 10.1109/TVT.2014.2361063. 53

[29]   M. Doumiati, A. C. Victorino and A. Charara, and D. Lechner. 2011. Onboard real-time estimation of vehicle lateral tire-road forces and sideslip angle. *IEEE/ASME Transactions Mechatronics*, 16(4), pp. 601–604. DOI: 10.1109/TMECH.2010.2048118. 53

[30]   J. Tjonnas and T. A. Johansen. 2010. Stabilization of automotive vehicles using active steering and adaptive brake control allocation. *IEEE Transactions on Control Systems Technology*, 18(3), pp. 545–558. DOI: 10.1109/TCST.2009.2023981. 53

[31]   W. Zhang and X. Guo. 2015. An ABS control strategy for commercial vehicle. *IEEE/ASME Transactions Mechatronics*, 20(1), pp. 384–392. DOI: 10.1109/TMECH.2014.2322629. 53

[32]   J. Ni and J. Hu. 2016. Handling performance control for hybrid 8-wheel-drive vehicle and simulation verification. *Vehicle System Dynamics*, 54(8), pp. 1098–1119. DOI: 10.1080/00423114.2016.1169303. 53

[33]   J. Ni, J. Hu, and C. Xiang. 2017. Relaxed static stability based on tire cornering stiffness estimation for all-wheel-drive electric vehicle. *Control Engineering Practice*, 64, pp. 102–110. DOI: 10.1016/j.conengprac.2017.04.011. 53

[34]   J. Ni, J. Hu, and C. Xiang. 2017. Envelope control for four-wheel-independently-actuated autonomous ground vehicle through AFS/DYC integrated control. *IEEE Transactions Vehicular Technology*. DOI: 10.1109/TVT.2017.2723418. 53

[35]   R. Castro, R. Araujo, M. Tanelli, S. M. Savaresi, and D Freitas. 2012. Torque blending and wheel slip control in EVs with in-wheel motors. *Vehicle System Dynamics*, 50, pp. 71–94. DOI: 10.1080/00423114.2012.666357. 53

[36]   V. Ivanov, D. Savitski, and B. Shyrokau. 2015. A survey of traction control and anti-lock braking systems of full electric vehicle with individually controlled electric mo-

tors. *IEEE Transactions Vehicular Technology*, 61(8), pp. 3394–3405. DOI: 10.1109/TVT.2014.2361860. 53

[37] Y. Chen and J. Wang. 2012. Design and evaluation on electric differentials for over-actuated electric ground vehicles with four independent in-wheel motors. *IEEE Transactions Vehicular Technology*, 61(4), pp. 1534–1542. DOI: 10.1109/TVT.2012.2187940. 53

[38] T. Hsiao. 2015. Robust wheel torque control for traction/braking force tracking under combined longitudinal and lateral motion. *IEEE Transactions Intelligent Transportation Systems*, 16(3), pp. 1335–1347. DOI: 10.1109/TITS.2014.2361515. 53

[39] Y. Shibahata, K. Shimada, and T. Tomari. 1993. Improvement of vehicle maneuverability by direct yaw moment control. *Vehicular System Dynamics*, 22, pp. 465–481. DOI: 10.1080/00423119308969044. 53

[40] M. Nagai, Y. Hirano, and S. Yamanaka. 1997. Integrated control of active rear wheel steering and direct yaw moment control. *Vehicular System Dynamics*, 27, pp. 357–370. DOI: 10.1080/00423119708969336. 54

[41] S. Sakai and Y. Hori. 1998. Robustified model matching control for motion control of electric vehicle. *IEEE AMC'98*. 54

[42] H. Zhang, G. Zhang, and J. Wang. 2016. H∞ observer design for LPV systems with uncertain measurements on scheduling variables: Application to an electric ground vehicle. *IEEE/ASME Transactions Mechatronics*, 21(3), pp. 1659–1670. DOI: 10.1109/TMECH.2016.2522759. 21, 54

[43] Z. Shuai, H. Zhang, J. Wang, J. Li, and M. Ouyang. 2014. Lateral motion control for four-wheel-independent-drive electric vehicles using optimal torque allocation and dynamic message priority scheduling. *Control Engineering Practice*, 24, pp. 55–66. DOI: 10.1016/j.conengprac.2013.11.012. 54

[44] Y. Chen, K. Hedrick and K. Guo. A novel direct yaw moment controller for in-wheel motor electric motors. *Vehicular System Dynamics*, 51(6), pp. 925–942, 2012. DOI: 10.1080/00423114.2013.773453. 54

[45] X. Huang, H. Zhang, G. Zhang, and J. Wang. 2014. Robust weighted gain scheduling H∞ vehicle lateral motion control with steering system backlash-type hysteresis. *IEEE Transactions on Control Systems Technology*, 22(5), pp. 1740–1753. DOI: 10.1109/TCST.2014.2317772. 54

[46] R. Wang, Y. Chen and D. Feng, X. Huang, and J. Wang. 2011. Development and performance characterization of an electric ground vehicle with independently actuated

in-wheel motors. *Journal Power Sources*, 196(8), pp. 3962–3971. DOI: 10.1016/j.jpowsour.2010.11.160. 54

[47]   K. Nam, H. Fujimoto, and Y. Hori. 2012. Lateral stability control of in-wheel-motor-driven electric vehicles based on sideslip angle estimation using lateral tire force sensors. *IEEE Transactions Vehicular Technology*, 61(5), pp. 1972–1984. DOI: 10.1109/TVT.2012.2191627. 54

[48]   S. Lee, K. Nakano, and Y. Hori. 2015. On-board identification of tire cornering stiffness using dual Kalman filter and GPS. *Vehicular System Dynamics*, 53(4), pp. 437–448. DOI: 10.1080/00423114.2014.999800. 54

[49]   K. Nam, H. Fujimoto, and Y. Hori. 2014. Advanced motion control of electric vehicles based on robust lateral tire force control via active front steering. *IEEE Transactions Vehicular Technology*, 19(1), pp. 289–299. DOI: 10.1109/TMECH.2012.2233210. 54

[50]   K. Nam, H. Fujimoto, and Y. Hori. 2013. Estimation of sideslip and roll angles of electric vehicles using lateral tire force sensors through RLS and Kalman filter approaches. *IEEE Transactions on Industrial Electronics*, 60(3), pp. 988–1000. DOI: 10.1109/TIE.2012.2188874. 54

[51]   L. Zhai, T. Sun, and J. Wang. 2016. Electronic stability control based on motor driving and braking torque distribution for a four in-wheel motor drive electric vehicle. *IEEE Transactions Vehicular Technology*, 65(6), pp. 4726–4739. DOI: 10.1109/TVT.2016.2526663. 54

[52]   Y. Wang, H. Fujimoto, and S. Hara. 2016. Torque distribution-based range extension control system for longitudinal motion of electric vehicles by LTI modeling with generalized frequency variable. *IEEE/ASME Transactions Mechatronics*, 21(1), pp. 443–452. DOI: 10.1109/TMECH.2015.2444651. 54

[53]   X. Yuan and J. Wang. 2012. Torque distribution strategy for a front-and rear wheel-driven electric vehicle. *IEEE Transactions Vehicular Technology*, 61(8), pp. 3365–3374. DOI: 10.1109/TVT.2012.2213282. 54

[54]   K. Maeda, H. Fujimoto, and Y. Hori. 2013. Four-wheel driving-force distribution. Method for instantaneous or split slippery roads for electric vehicle. *Automatica*, 54(1), pp. 103–113. DOI: 10.7305/automatika.54-1.312. 54

[55]   H. Du, N. Zhang, and G. Dong. 2010. Stabilizing vehicle lateral dynamics with considerations of parameter uncertainties and control saturation through robust yaw control. *IEEE Transactions Vehicular Technology*, 59(5), pp. 2593–2597. DOI: 10.1109/TVT.2010.2045520. 54

[56]   H. Zhang, X. Zhang, and J. Wang. 2014. Robust gain scheduling energy to peak control of vehicle lateral dynamics stabilization. *Vehicular System Dynamics*, 52(3), pp. 309–340. DOI: 10.1080/00423114.2013.879190. 54

[57]   X. Huang, H. Zhang, G. Zhang, and J. Wang. 2014. Robust weighted gains scheduling H∞ vehicle lateral motion control with steering system backlash-type hysteresis. *IEEE Transactions on Control Systems Technology*, 22(5), pp. 1740–1753. DOI: 10.1109/TCST.2014.2317772. 54

[58]   Z. Shuai, H. Zhang, J. Wang, J. Li, and M. Ouyang. 2014. Combined AFS and DYC control of four-wheel-indepdent-drive electric vehicles over CAN network with time-varying delays. *IEEE Transactions Vehicular Technology*, 63(2), pp.591–602. DOI: 10.1109/TVT.2013.2279843. 54

[59]   H. Zhang and J. Wang. 2016. Vehicle lateral dynamics control through AFS/DYC and robust gain-scheduling approach. *IEEE Transactions Vehicular Technology*, 65(1), pp. 489–494. DOI: 10.1109/TVT.2015.2391184. 21, 54

[60]   R. Marino and F. Cinili. 2009. Input-output decoupling control by measurement feedback in four-wheel-steering vehicles. *IEEE Transactions on Control Systems Technology*, 17(5), pp. 1163–1170. DOI: 10.1109/TCST.2008.2004441. 54

[61]   T. Lam, H. Qian, and Y. Xu. 2010. Omnidirectional steering interface and control for a four-wheel independent steering vehicle. *IEEE/ASME Transactions Mechatronics*, 15(3), pp. 329-338. DOI: 10.1109/TMECH.2009.2024938. 54

[62]   H. Russell and J. Gerdes. 2016. Design of variable vehicle handling characteristics using four-wheel steer-by-wire. *IEEE Transactions on Control Systems Technology*, 24(5), pp. 1529–1540. DOI: 10.1109/TCST.2015.2498134. 54

[63]   M. Li and Y. Jia. 2014. Decoupling control in velocity-varying four-wheel steering vehicles with H-infinity performance longitudinal velocity and yaw rate feedback. *Vehicular System Dynamics*, 52(12), pp. 1563–1583. DOI: 10.1080/00423114.2014.951060. 54

[64]   F. Fahimi. 2013. Full drive-by-wire dynamic control for four-wheel-steer all-wheel-drive vehicles. *Vehicular System Dynamics*, 51(3), pp. 360–376. DOI: 10.1080/00423114.2012.743668. 54

[65]   Y. Song, H. Chen, and D. Li. 2011. Virtual-point-based fault-tolerant lateral and longitudinal control of 4WS vehicles. *IEEE Transactions Intelligent Transportation Systems*, 12(4), pp. 1343–1351. DOI: 10.1109/TITS.2011.2158646. 54

[66]   M. Li, Y. Jia, and J. Du. 2014. LPV control with decoupling performance of 4WS vehicles under velocity-varying motion. *IEEE Transactions on Control Systems Technology*, 22(5), pp. 1708–1724. DOI: 10.1109/TCST.2014.2298893. 54

[67]   R. Wang, C. Hu, and Z. Wang. 2015. Integrated optimal dynamics control of 4WD4WS electric ground vehicle with tire road frictional coefficient estimation. *Mechanical Systems Signal Processing*, 60, pp. 728–741. DOI: 10.1016/j.ymssp.2014.12.026. 54

[68]   C. Hu, R. Wang, F. Yan, and N. Chen. 2016. Output constraint control on path following of four-wheel independently actuated autonomous ground vehicle. *IEEE Transactions Vehicular Technology*, 65(6), pp. 4033–4043. DOI: 10.1109/TVT.2015.2472975. 54

[69]   P. Song, M. Tomizuka, and C. Zong. 2015. A novel integrated chassis controller for full drive-by-wire vehicles. *Vehicular System Dynamics*, 53(2), pp. 215–236. DOI: 10.1080/00423114.2014.991331. 54

[70]   X. Huang, H. Zhang, G. Zhang, and J. Wang. 2014. Robust weighted gain scheduling H∞ vehicle lateral motion control with steering system backlash-type hysteresis. *IEEE Transactions on Control Systems Technology*, 22(5), pp. 1740–1753. DOI: 10.1109/TCST.2014.2317772. 55

[71]   S. Erlien, S. Fujita, and J. C. Gerdes. 2016. Shared steering control using safe envelopes for obstacle avoidance and vehicle stability. *IEEE Transactions Intelligent Transportation Systems*, 17(2), pp. 441–451. DOI: 10.1109/TITS.2015.2453404. 55

[72]   R. Wang, C. Hu, F. Yan, and M. Chadli. 2016. Composite nonlinear feedback control for path following of four-wheel independently actuated autonomous ground vehicle. *IEEE Transactions Intelligent Transportation Systems*, 17(7), pp. 2063–2074. DOI: 10.1109/TITS.2015.2498172. 55

[73]   R. Wang, H. Jing, C. Hu, F. Yan, and N. Chen. 2016. Robust H∞ path following control for autonomous ground vehicles with delay and data dropout. *IEEE Transactions Intelligent Transportation Systems*, 17(7), pp. 2042–2049. DOI: 10.1109/TITS.2015.2498157. 55

[74]   X. Zhu, H. Zhang, and J. Wang. 2015. Robust lateral motion control of electric ground vehicles with random network-induced delays. *IEEE Transactions Vehicular Technology*, 64(11), pp. 4985–4955. DOI: 10.1109/TVT.2014.2383402. 55

[75]   H. Her, Y. Koh, E. Joa, K. Yi, and K. Kim. 2016. An integrated control of differential braking, front/rear traction, and active roll moment for limit handling performance. *IEEE Transactions Vehicular Technology*, 65(6), pp. 4288–4300. DOI: 10.1109/TVT.2015.2513063. 55

[76]   J. Ni and J. Hu. 2017. Dynamics control of autonomous vehicle at driving limits and experiment on an autonomous Formula racing car. *Mechanical Systems and Signal Processing*, 90, pp. 154–174. DOI: 10.1016/j.ymssp.2016.12.017. 55

[77]   M. R. Anderson and W. H. Mason. 1996. An MDO approach to Control-configured-vehicle design. *Proceedings of American Institute of Aeronautics and Astronautics Conference*, pp. 734–743. DOI: 10.2514/6.1996-4058. 55

[78]   H. G. Kwatny, W. H. Bennett, and J. Berg. 1991. Regulation of relaxed static stability aircraft. *IEEE Transactions on Automatic Control*, 36(11), pp. 1315–1323. DOI: 10.1109/9.100946. 55

[79]   J. Ni, J. Hu, and C. Xiang. Control-configured-vehicle design and implementation on x-by-wire electric vehicle. *IEEE Transactions Vehicular Technology* (Under Review).57, 61

[80]   J. Ni, J. Hu, and C. Xiang. Relaxed static stability for four-wheel independently actuated ground vehicle. *IEEE/ASME Transactions Mechatronics* (Under Review). 73

Author Biographies

Jun NI received his B.E. in mechanical engineering from Beijing Institute of Technology, Beijing, China, in 2013. He is currently a Ph.D. candidate with the school of Mechanical Engineering at Beijing Institute of Technology. He is also a visiting Ph.D. student in Vehicle Dynamics and Control Lab at University of California, Berkeley.

Jun's research interest is the intelligent control for the UGV or mobile robots. He has published over 40 papers and holds 10 Chinese patents in this field. He served as the President of the BIT Special UGV Innovation Center and BIT Formula Student Racecar team. He is the co-organizer of several international conferences, such as the special workshop of the IEEE Intelligent Vehicle Symposium and the special workshop of the 19th APAC SAE Congress. Jun won the National Award of Science and Technology for Youth in 2013, which was awarded by the Vice President of China. He was selected into the National Support Program for Youth Talent in 2016, which was awarded by the China Association for Science and Technology.

Jibin Hu received his B.E. in mechanical engineering from Beijing Institute of Technology, Beijing, China in 1992, and his Ph.D. from Beijing Institute of Technology in 2003. He is currently a professor at Beijing Institute of Technology. He currently serves as the Vice Dean of School of Mechanical Engineering at Beijing Institute of Technology.

Jibin's s research interest includes vehicle transmission system design and control, vehicle dynamics and control, as well as autonomous vehicle control. He has published over 80 papers and got over 20 Chinese and international patents in these fields. He is currently serving as the vice director of the Chinese National Vehicle Transmission System Key Lab. He is also served as the advisor of the BIT Formula Student Racecar team. He has been the co-organizer of many international conferences, including the IEEE Transportation Electrification Conference of ASIA-PACIFIC. He has been supported by the program for New Century Excellent Talents in Chinese Universities and has won the National Science and Technology Reward.

Changle Xiang received his Ph.D. in Beijing Institute of Technology in 2002. He is a currently a professor at Beijing Institute of Technology. He serves as the Vice President of Beijing Institute of Technology.

Changle's research interests include vehicle transmission system design and control, vehicle dynamics and control, as well as autonomous vehicle control. He is a famous scholar in this field in China, and has published over 100 papers and holds over 40 Chinese and international patents. His research efforts have been widely used in many types of the automobiles in China, especially military vehicles. He is currently serving as the director of the National Vehicle Transmission System Key Lab and the Autonomous Vehicle Platform Key Lab of Ministry of Industrialization and Information Technology. He has served as the organizer of many international conferences, and the director of the Off-Road Vehicle Branch of SAE China. He has been elected to the Changjiang Scholar since 2004 and has won the National Science and Technology Reward, National Technology Invention Reward multiple times.

Printed in the United States
by Baker & Taylor Publisher Services